T0312955

JPEG Series

RIVER PUBLISHERS SERIES IN SIGNAL, IMAGE AND SPEECH PROCESSING

Series Editors

MARCELO SAMPAIO DE ALENCAR
Universidade Federal da Bahia UFBA
Brasil

MONCEF GABBOUJ
Tampere University of Technology
Finland

THANOS STOURAITIS
University of Patras
Greece
and
Khalifa University
UAE

Indexing: all books published in this series are submitted to the Web of Science Book Citation Index (BkCI), to SCOPUS, to CrossRef and to Google Scholar for evaluation and indexing

The "River Publishers Series in Signal, Image and Speech Processing" is a series of comprehensive academic and professional books which focus on all aspects of the theory and practice of signal processing. Books published in the series include research monographs, edited volumes, handbooks and textbooks. The books provide professionals, researchers, educators, and advanced students in the field with an invaluable insight into the latest research and developments.

Topics covered in the series include, but are by no means restricted to the following:

- Signal Processing Systems
- Digital Signal Processing
- Image Processing
- Signal Theory
- Stochastic Processes
- Detection and Estimation
- Pattern Recognition
- Optical Signal Processing
- Multi-dimensional Signal Processing
- Communication Signal Processing
- Biomedical Signal Processing
- Acoustic and Vibration Signal Processing
- Data Processing
- Remote Sensing
- Signal Processing Technology
- Speech Processing
- Radar Signal Processing

For a list of other books in this series, visit www.riverpublishers.com

JPEG Series

K.R. Rao

The University of Texas at Arlington,
Department of Electrical Engineering Arlington, USA

Humberto Ochoa Domínguez

Universidad Autónoma de Ciudad Juárez, México

Shreyanka Subbarayappa

Ramaiah University of Applied Sciences, India

LONDON AND NEW YORK

Published 2021 by River Publishers
River Publishers
Alsbjergvej 10, 9260 Gistrup, Denmark
www.riverpublishers.com

Distributed exclusively by Routledge
4 Park Square, Milton Park, Abingdon, Oxon OX14 4RN
605 Third Avenue, New York, NY 10158

First published in paperback 2024

JPEG Series / by K.R. Rao, Humberto Ochoa Domínguez, Shreyanka Subbarayappa.

Routledge is an imprint of the Taylor & Francis Group, an informa business

Publisher's Note
The publisher has gone to great lengths to ensure the quality of this reprint but points out that some imperfections in the original copies may be apparent.

While every effort is made to provide dependable information, the publisher, authors, and editors cannot be held responsible for any errors or omissions.

ISBN: 978-87-7022-593-9 (hbk)
ISBN: 978-87-7004-309-0 (pbk)
ISBN: 978-1-003-33873-4 (ebk)

DOI: 10.1201/9781003338734

Contents

Preface   xiii

Foreword   xv

Acknowledgement   xvii

List of Figures   xix

List of Tables   xxv

List of Abbreviations   xxvii

1   Introduction   1
    1.1   JPEG Series   2

2   Digital images   7
    2.1   Introduction   7
    2.2   One-Dimensional Sampling   8
    2.3   Reconstruction   13
    2.4   Sampling of Continuous-Time Two-Dimensional
        Functions   17
    2.5   Image Reconstruction from Its Samples   21
    2.6   Image Quantization   23
        2.6.1   The uniform quantizer   25
        2.6.2   Quantization error   26
        2.6.3   The Optimal Quantizer (Lloyd–Max Quantizer) . . .   28
        2.6.4   The Mid-Tread Quantizer   31
    2.7   The Human Visual System   32
        2.7.1   HVS-Based Quantization   33
    2.8   Color Space   36
        2.8.1   YCbCr Formats   45
    2.9   Chromaticity Diagram   45

3   The Discrete Cosine and Sine Transforms   **49**
   3.1   The DCT   50
   3.2   The DST   53
   3.3   The Fast DCT   54
   3.4   Structures for Implementing the Fast 1-D DCT   58
   3.5   2D DCT   60
   3.6   Hierarchical Lapped Transform   62
   3.7   Problems   65

4   Entropy Coding   **67**
   4.1   Source Entropy   67
   4.2   Run-Length Coder   68
   4.3   Huffman Coder   69
   4.4   Arithmetic Coder   71
          4.4.1   Binary Arithmetic Coding (BAC)   74
   4.5   Problems   80

5   JPEG   **85**
   5.1   JPEG Parts   85
   5.2   Image Pre-Processing   86
   5.3   Encoder   87
   5.4   Modes of Operation   96
   5.5   Some Limitations of JPEG Encoder 102
   5.6   Table of Metadata Structure 103

6   JPEG XR   **105**
   6.1   JPEG XR Parts 106
   6.2   Image Pre-Processing 108
          6.2.1   Pre-Scaling 108
          6.2.2   Color Conversion 108
   6.3   Encoder 112
          6.3.1   Transform 112
          6.3.2   Steps of transformation 113
          6.3.3   Quantization 115
                   Quantization control on a spatial region basis: 116
                   Quantization control on a frequency band basis: . . . 117
                   Quantization control on a color plane
                   component basis: 117
                   Quantization control type combinations: 117

6.3.4   Prediction of Transform Coefficients and
Coded Block Patterns 117
6.3.5   Adaptive Scanning Pattern 119
6.3.6   Adaptive Coefficient
Normalization (Entropy Coding) 120
6.4   Codestream Structure 121
6.5   Profiles 122
6.6   Levels 123
6.7   Decoder 125
6.7.1   Parsing Process 126
6.7.2   Decoding Process 126
Sample reconstruction: 126
6.8   Motion JPEG XR 127
6.8.1   Sample Entry and Sample Format 128
6.8.2   Syntax 128
6.8.3   Semantics 130
6.8.4   Profiles and Levels 131
Advanced profile "mjxr": 131
Sub-Baseline profile 'mjxs' 131
6.9   Summary 132

7   JPEG XT   **135**
7.1   JPEG XT Parts 135
7.2   High Dynamic Range 136
7.3   Encoder 137
7.3.1   Profile A 138
7.3.2   Profile B 139
7.3.3   Profile C 139
7.3.4   Profile D 140
7.4   Lossless Coding and Lifting 141
7.5   Coding of Opacity Information 142
7.6   Privacy and Protection 143
7.7   Decoder 143
7.8   Summary 147

8   JPEG 2000   **149**
8.1   JPEG 2000 Parts 150
8.2   Architecture of JPEG-2000 152
8.3   Reference Grid 152
8.4   Tiling Grid and Tiles 154

8.5   Image Pre-Processing 156
8.6   Forward Intercomponent Transform. 156
8.7   Forward Intracomponent Transform. 157
    8.7.1   The Continuous Wavelet Transform. 157
    8.7.2   Multiresolution Analysis 159
    8.7.3   The Discrete Wavelet Transform (DWT) 162
    8.7.4   1D Filter Bank. 162
    8.7.5   2D Filter Bank. 164
    8.7.6   Two-Band Analysis and Synthesis Filter Bank 166
        z-transform of the analysis bank: 166
        z-transform of the synthesis bank:. 167
    8.7.7   Biorthogonality 168
    8.7.8   Irreversible and Reversible Wavelet Filters 170
    8.7.9   Convolution- and Lifting-Based Filters 171
8.8   Quantization 173
8.9   Partitioning of Subbands 174
8.10   Entropy Coding 175
8.11   Rate Control. 180
8.12   Region of Interest (ROI) Coding 182
8.13   Error Resilience. 182
8.14   JPEG 2000 Part 2 182
8.15   Motion JPEG 2000 183
8.16   Volumetric Data Encoding 183
8.17   Some Limitations of JPEG 2000 185
8.18   Summary 186

9   JPEG XS   **189**
9.1   JPEG XS Parts. 189
9.2   Architecture of JPEG XS 190
9.3   Multi-Component Transform 191
9.4   Forward Transform 191
9.5   Pre-Quantization. 192
9.6   Quantization 192
9.7   Rate Control. 193
9.8   Entropy Coding 194
9.9   Subjective Evaluations and Setup 196
9.10   Test Material. 199
    9.10.1   VQEG 199
        Sequences S01, S02:. 199

9.10.2 ARRI 199
     Sequences S03, S04:. 199
     Sequences S05, S06:. 199
9.10.3 Blender 199
     Sequences S07, S08 (Sintel): 199
     Sequence S09 (Tears of steel): 200
9.10.4 EBU 200
     Sequence S12 (PendulusWide) 200
     Sequence S13 (RainFruits) 200
9.10.5 Apple 200
     Sequence S14:. 200
9.10.6 Huawei. 200
     Sequence S15:. 200
9.10.7 Richter. 201
     Sequence S16:. 201
9.10.8 Copyright Condition Statements for
     Test Still Images. 201
     Zoneplates images: 201
     HorseFly: 201
9.10.9 Tools. 201
     Peacock:. 202
     HintergrundMusik: 202
     LeavesIso200:. 202
9.10.10 Acknowledgments 202

10 JPEG Pleno   **203**
10.1 Plenoptic Representation 203
10.2 JPEG Pleno Parts 206
10.3 Holographic Imaging 206
10.4 Point Cloud Imaging 208
10.5 Light-Field Imaging 208
10.6 JPEG Pleno Test Conditions. 209
10.6.1 Holographic Data 210
     Holographic specific requirements: 212
     Bit rates and rate metrics: 214
     Quality metrics: 214
         SNR and PSNR 214
         SSIM. 215
         Bjøntegaard metric 215

VIFp 215
SNR of first-order wavefield. 216
RMSE of retrieved phase 216
Hamming distance. 216
Color information metric 217
Configuration quality metrics and anchor codecs: . . 217
10.6.2 Point Cloud 218
Test set material:. 218
Subjective quality evaluation:. 220
Considerations for subjective evaluation: 221
Rate and rate-distortion evaluation: 222
Scalability:. 222
Random access: 222
Quality metrics: 223
Point-to-geometry D1 metric 223
Point-to-plane geometry D2 metric 224
Point-to-plane angular similarity metric. . . . 224
Point-to-point attribute measure. 225
Target bit rates 225
Anchor generation: 225
Anchor codecs 225
10.6.3 Light Field. 226
Test set material:. 226
Lenslet Lytro Illum Camera: 226
Fraunhofer High Density
Camera Array (HDCA): 226
Poznan HDCA: 226
Stanford HDCA:. 226
Synthetic HCI HDCA:. 226
Quality metrics: 227
RGB to YCbCr conversion 227
PSNR 228
SSIM. 228
Bjøntegaard metric 228
Rate metrics:. 228
Random access metric: 228
HEVC anchor generation:. 229

11 JPEG AIC   231

  11.1 Advanced Image Coding 231
  11.2 AIC Characteristics 232
  10.3 AIC Codec 232

12 JPEG LS   235

  12.1 JPEG LS Parts 235
  12.2 JPEG-LS Encoder Architecture 236
    12.2.1 JPEG-LS Modeler 237
      Context: 237
      Predictor: 238
    12.2.2 JPEG-LS Coder 241
      Run mode: 241
      Coding parameters: 242
      Codec initialization: 243
      Encoding procedure for single component: 243
    12.2.3 Encoding Procedure for Single Component 244
  12.3 Marker Assignment 253
    12.3.1 Decoding Procedure 253

13 JPEG XL   257

  13.1 Scope 258
  13.2 Use Cases 258
  13.3 Requirements 259
  13.4 Compressed Bit-Stream Requirements 259
  13.5 Test Material 260
  13.6 Anchors 260
  13.7 Target Rates 261
  13.8 Objective Quality Testing 261
  13.9 Subjective Quality Testing 261
  13.10 Evaluation Tools 261

14 JPSearch   263

  14.1 JPSearch Parts 264
    14.1.1 Part 1: Global Architecture 264
    14.1.2 Part 2: Schema and Ontology 266
    14.1.3 Part 3: Query Format 270
    14.1.4 Part 4: File Format 271
    14.1.5 Part 5: Data Interchange Format 273
    14.1.6 Part 6: Reference Software 274

15 JPEG Systems   **277**
    15.1 JPEG Systems Parts   277

16 JBIG   **279**
    16.1 Subject Matter for JBIG Coding   280
    16.2 Relationship Between Segments and Documents   280
    16.3 JBIG2 Decoder   282
        16.3.1 Decoding Procedure   284
    16.4 Lossy Coding   285
        16.4.1 Symbol Coding   285
        16.4.2 Generic Coding   286
        16.4.3 Halftone Coding   286
    16.5 Summary   287

ANNEX A. PROVISION   **289**
    PROVISION – PeRceptually Optimized VIdeo compresSION . . .   289

References   **293**

Index   **311**

About the Authors   **315**

Preface

Image coding standards provide interoperability between codecs built by different manufacturers and provide state-of-the-art technology developed by a group of experts in this field. The main goal of the continuous efforts on image coding standardization is to achieve low bit rate for data storage and transmission, feasibility of hardware implementation while maintaining acceptable distortion.

Different approaches have been utilized. Some of them use the discrete cosine transform (DCT) as their main engine, like the JPEG, whose aim was to develop a general method of compression for a wide range of applications for still images. The discrete wavelet transforms (DWT) that provide the core technology for the JPEG 2000 still image coding standard intended to yield superior compression ratios and image quality than JPEG.

With the rapid developments of imaging technology, image compression, coding tools and techniques, it is necessary to evolve coding standards to provide compatibility and interoperability between the image communication and storage products manufactured by different vendors.

This book is a collection of the current standards used for still image coding and provides a comprehensive overview of the standards developed and under development by the International Telecommunication Union (ITU), the International Organization for Standardization (ISO), and the International Electrotechnical Commission (IEC).

The book describes the JPEG standard, which is the most common format used by the digital cameras and other devices; the JPEG XR, that supports a wide range of color encoding formats and n-component encoding, with a variety of bit depths for a wide range of applications especially for high dynamic range imagery; the JPEG XT, which provides compression of high-dynamic-range imaging and representation of alpha channels; the JPEG 2000 which is based on the DWT to provide an extremely high level of scalability and accessibility; the JPEG XS that specifies a compression technique with end-to-end low latency and low complexity; the JPEG Pleno, that is a framework for representing new imaging modalities, such as point-cloud, light-field, and holographic. The aim is to provide the system tools, coding tools, and metadata for data and metadata manipulation, editing, random access and interaction, protection of privacy, ownership rights and security management; the JPEG AIC that locates and evaluates new scientific developments and advancements in image coding research; the JPEG-LS, for effective lossless and near lossless applications; the JPEG XL, optimized for responsive web environments; the JPSearch, that is an initiative of the JPEG Committee to address the management and exchange of metadata of images. the JPEG Systems, that defines an overall framework for future and legacy standards; and finally, the JBIG standard developed for bi-level images.

This work is intended to be a reference for engineers and researchers in the fields of image compression as a good reference for the JPEG series standards.

Foreword

While JPEG has been developed more than 25 years ago, it is still dominant in terms of image formation, manipulation, and transmission over the Internet and other media. On the other hand, as technology advances, new demands have arisen for the efficient transmission and storage of images causing other formats to emerge. Over the years, several extensions such as, JPEG LS, JPEG AIC, JPEG 2000, JPEG XT, JPEG XR, JPEG XS, JPEG XL, JPEG Search, JPEG Systems, and JPEG Pleno have been added, constructing the collection or series of standards for the compression and transmission of images. However, as far as we know, there has not been a collection that brings together most of the series of JPEG standards in a book. This book covers the descriptions of the JPEG standards and gives the reader an overview of the latest advances in the standards of the Joint Photographic Experts Group.

.

Acknowledgements

Dr. Rao would like to acknowledge the support provided by the University of Texas at Arlington (UTA), the graduate students and alumnae in the Multimedia Processing Lab (MPL) at UTA in various ways have made constructive comments.

Dr. Ochoa would like to express his gratitude to the Department of Ingeniería Eléctrica y Computación of the Universidad Autónoma de Ciudad Juárez (UACJ). Especial thanks to La Cordi Cynthia, Isela, Ismael and the graduate students in the Signal Processing Lab for their continuous support which made possible the realization of this book.

Dr. Shreyanka wishes to thank her son Ashrith A. and her husband Adarsh K. for understanding and encouraging her during the process of this book write up. She would also like to mention her only inspirations, her parents, Prof. H. Subbarayappa and Anasuya Subbarayappa for their continuous support and unconditional help without which this would not have been possible.

List of Figures

Figure 2.1   Representation of (a) the original image,
(b) the image of 64 × 96 pixels, and
(c) image of 128 × 192 pixels.   8

Figure 2.2   Discrete representation of 1D continuous-time
signals.   8

Figure 2.3   Diagram of the ideal continuous to discrete
converter (Oppenheim, 2014).   9

Figure 2.4   Frequency domain of the sampling process.
Spectrum of (a) the original signal or baseband
signal,
(b) the sampling function, and
(c) the sampled signal.   10

Figure 2.5   Diagram of the ideal reconstruction
system (Oppenheim, 2014)..   14

Figure 2.6   Image corresponding to the function
$f(v,w) = 20 - 4w$.   17

Figure 2.7   Image with length W and height H, both in
units of length.   18

Figure 2.8   Sampling grid.   19

Figure 2.9   Fourier transform of (a) a band-limited
2D function and (b) its region of support.   20

Figure 2.10   Replicas of the sampled image spectrum
with no aliasing..   22

Figure 2.11   Midrise quantizer..   24

Figure 2.12   Uniform quantizer transfer function for
$B = 2$, $L = 4$ in the midrise quantizer.   26

Figure 2.13   Images (a) original, (b) quantized using 2 bits or
4 gray levels, MSE = 337.13, and (c) quantized
using 3 bits or 8 gray levels, MSE = 92.79..   28

Figure 2.14   Histogram of frequencies of the gray-scale of the original image in Figure 2.13(a). The vertical lines denote the boundaries of the decision levels and the numbers inside of each bin of the reconstruction levels of a 2-bit uniform optimal quantizer.   29

Figure 2.15   Histogram of frequencies of the gray-scale of the original image in Figure 2.13(a). The vertical lines denote the boundaries of the decision levels and the numbers inside of each bin of the reconstruction levels of a 3-bit uniform optimal quantizer.   29

Figure 2.16   Comparison of (a) midrise and (b) midtread quantizers..   31

Figure 2.17   Contrast sensitivity functions of luminance signals, and red-green and blue-yellow opponent chromatic signals (McKeefry, Murray, & Kulikowski, 2001) (dash lines are extrapolated experimental data). . . .   33

Figure 2.18   Human visual frequency weighting of DCT coefficients. 35

Figure 2.19   Cartesian coordinate system of RGB color space. . .   37

Figure 2.20   RGB−HSI relation.   38

Figure 2.21   Intensity and saturation of a point P in the RGB space.   39

Figure 2.22   Projection perpendicular to the intensity axis.   40

Figure 2.23   Projection of shaded plane in Figure 2.20.   40

Figure 2.24   Positioning of YCbCr samples for the (a) 4:4:4, (b) 4:2:2, (c) 4:1:1 co-sited sampling, (d) 4:2:0 for H.261, H.263 MPEG 1, and (e) 4:2:0 for MPEG 2 sampling formats..   46

Figure 2.25   CIE 1931 chromaticity diagram.   47

Figure 3.1   Basis function waveforms of the 1D-DCT-II..   52

Figure 3.2   Basis function waveforms of the 1D-DST-II..   54

Figure 3.3   Signal flowgraph of the 8-point, forward 1D-DCT of Vetterli and Ligtenberg.   58

Figure 3.4   Signal flowgraph of the 8-point, fast (scaled) AAN forward 1D-DCT.   59

Figure 3.5   Signal flowgraph for computing DCT via DHT. $C_k = \cos(k\pi/32)$ and $S_k = \sin(k\pi/32)$, $N = 16$. . .   59

Figure 3.6   Basis images for (a) 4×4 size and (b) 8×8 size. . .   62

Figure 3.7 Lattice structure of the four-channel hierarchical lapped transform. 63

Figure 4.1 Arithmetic coding process with intervals scaled up at each stage. 72

Figure 5.1 General block diagram of the JPEG encoder. 87

Figure 5.2 Two 8 × 8 transform blocks showing 64 coefficients each. The DC coefficient is black and the remaining 63 AC components are the white squares. 88

Figure 5.3 Zig-zag order. 90

Figure 5.4 Huffman entropy encoder. 95

Figure 5.5 Arithmetic entropy encoder. 95

Figure 5.6 JPEG encoder block diagram. 96

Figure 5.7 Block diagram of sequential JPEG encoder and decoder. 97

Figure 5.8 Display of an image using JPEG baseline. 98

Figure 5.9 Block of 8 × 8 DCT coefficients. (a) Spectral bands. (b) Zig-zag scan. 98

Figure 5.10 Progressive JPEG. 99

Figure 5.11. Neighboring samples (*a*, *b*, and *c* = reconstructed samples; *x* = sample under prediction). 100

Figure 5.12 Hierarchical images resulting after hierarchical mode. 100

Figure 5.13 A three-level hierarchical encoder. 101

Figure 6.1 Block diagram of JPEG XR (a) encoder and (b) decoder (ITU-T-Rec-T.832 | ISO/IEC-29199-2, 2009). 107

Figure 6.2 Overview of image partitions and internal windowing. 111

Figure 6.3 Example of image partitioning in JPEG XR. 112

Figure 6.4 Frequency hierarchy of a macroblock. (a) Luma and full resolution chroma. (b) YUV 422. (c) YUV420. 114

Figure 6.5 Two-stage lapped biorthogonal transform. 116

Figure 6.6 DC prediction (Srinivasan *et al.*, 2007). 118

Figure 6.7 LP coefficients (Srinivasan *et al.*, 2007). 118

Figure 6.8 HP left prediction scheme (Srinivasan *et al.*, 2007). 119

Figure 6.9 Codestream structure (ISO/IEC-29199, 2010). . . . 121

Figure 6.10 Informative decoding process block diagram. 125

Figure 7.1   JPEG XT profile A encoder.   138

Figure 7.2   JPEG XT profile B encoder.   139

Figure 7.3   JPEG XT profile C encoder.   140

Figure 7.4   Opacity-enabled JPEG XT decoder.   142

Figure 7.5   Organization of the APP11 marker segment
for boxes.   144

Figure 7.6   The simplified JPEG XT standard decoder
architecture. The gray block is the legacy decoding
system (JPEG).   145

Figure 8.1   JPEG 2000 codec. (a) Encoder. (b) Decoder.   153

Figure 8.2   Reference grid.   153

Figure 8.3   Reference grid, tiling grid, and image area..   155

Figure 8.4   Tile-component coordinate system..   156

Figure 8.5   Nested function spaces spanned by (a) the scaling
function and (b) relationship between the scaling
and wavelet function spaces, where $\oplus$ indicates
union of subspaces.   161

Figure 8.6   One-stage, 1D analysis of DWT coefficients..   163

Figure 8.7   Two-stage, 1D analysis of DWT coefficients.   164

Figure 8.8   One-stage, 1D synthesis of DWT coefficients.   164

Figure 8.9   One-stage, 2D analysis of DWT coefficients..   165

Figure 8.10   One level of 2D-DWT decomposition. Upper-
left subbands are the approximation coefficients,
upper-right subbands are the horizontal detail
coefficients, lower-left subbands are the vertical
detail coefficients, and lower-right subbands are
the diagonal detail coefficients..   165

Figure 8.11   Two levels of 2D-DWT decomposition.   165

Figure 8.12   Two-band analysis filter bank.   166

Figure 8.13   Lifting realization of one-stage, 1D, two-channel
perfect reconstruction uniformly maximally
decimated filter bank (UMDFB). (a) Analysis side.
(b) Synthesis side..   171

Figure 8.14   Precincts, packets, and code-blocks.   175

Figure 8.15   Division of subbands into code-blocks of same
dimensions in every subband..   176

Figure 8.16   Simple concatenation of optimally truncated code-
block bit stream.   177

Figure 8.17   Quality layers. Numbers inside a layer indicate
sequence of code-block contributions required for
a quality progressive stream.   178

Figure 8.18   Scan pattern for bitplane coding within a
code-block.   179

Figure 8.19   Partitioning of volumetric data in volumetric
tiles (left) and subband decomposition (right)
(Bruylants, Munteanu, Alecu, Decklerk, &
Schelkens, 2007).   184

Figure 8.20   3D code-block scanning pattern (Bruylants,
Munteanu, Alecu, Decklerk, & Schelkens, 2007)
showing the scanning of two 3D $8 \times 8 \times 4$ code-
block consisting of four 8×8 slices.   185

Figure 9.1   Architecture of the JPEG XS (a) encoder and
(b) decoder.   191

Figure 10.1   Parameterization of one ray in 3D by
position (x, y, z) and direction (θ, Φ).   204

Figure 10.2   Plenoptic function describes all the images visible
from a particular position.   205

Figure 10.3   Holographic data (Xing, 2015)..   207

Figure 10.4   JPEG Pleno light fields codec architecture
(Ebrahimi, 2019).   209

Figure 10.5   Encoding/decoding pipeline
(ISO/IECJTC1/SC29/WG1/N84025, 2019).   229

Figure 11.1   AIC codec (Van Bilsen, 2018).   233

Figure 11.2   Prediction modes used by the AIC codec.   234

Figure 12.1   JPEG-LS block diagram.   237

Figure 12.2   Division-free for bias computation procedure.   240

Figure 12.3   LSE marker (a) preset parameters marker segment
syntax and (b) preset coding parameters.   242

Figure 12.4   Example of sampling for a three-component image. .   252

Figure 12.5   Example line interleaving for a three-component
image.   253

Figure 14.1   JPSearch system global architecture.   265

Figure 14.2   JPEG or JPEG 2000 file formats, with the potential
to have multiple JPSearch files..   266

Figure 14.3   Example of image annotation.   269

Figure 14.4   Example of the description of a fictional image
(JPSearch-RA, 2011)..   270

Figure 14.5   Structure of JPEG compatible JPSearch file format. .   271

Figure 14.6   Structure of JPEG 2000 compatible JPSearch file
format..   272

Figure 14.7   JPSearch reference software architecture.   275

Figure 16.1   JBIG2 decoder block diagram.   283

List of Tables

Table 2.1   Typical bandwidths and sampling frequencies. 12
Table 2.2   Decision and reconstruction levels for $B = 2$. 26
Table 2.3   Decision and reconstruction levels for $B = 3$. 27
Table 2.4   The human visual frequency-weighting
            matrix $H(v,w)$ (Long- Wen, Ching-Yang, &
            Shiuh-Ming, 1999). 36
Table 2.5   HVS-based luminance quantization table $Q_{HS}(v,w)$
            (Long-Wen, Ching-Yang, & Shiuh-Ming, 1999). . . . 36
Table 5.1   Luminance quantization table. 89
Table 5.2   Chrominance quantization table. 89
Table 5.3   Scaled luminance quantization table ($Q = 80$). 90
Table 5.4   Scaled chrominance quantization table ($Q = 80$). . . . 90
Table 5.5   Coefficient encoding category. 91
Table 5.6   Suggested Huffman code for luminance
            DC differences. 91
Table 5.7   Suggested Huffman code for chrominance
            DC difference. 92
Table 5.8   Partial Huffman table for luminance for
            AC Run/Size pairs. 93
Table 5.9   Partial Huffman table for chrominance for
            AC Run/Size pairs. 94
Table 5.10 Predictors for lossless coding for a given scan
            though a component or group of components. 99
Table 5.11 The common JPEG markers. 103
Table 5.12 Example of segments of an image coded using
            baseline DCT. 104
Table 6.1   JPEG XR profiles (ISO/IEC-29199, 2010),
            (Rec. ITU-T, 2016). 122
Table 6.2   Levels of JEPG XR (JPEG, Overview of
            JPEG XR, 2018), (Rec. ITU-T, 2016). 123
Table 8.1   Daubechies 9/7 analysis filter coefficients. 170
Table 8.2   Daubechies 9/7 synthesis filter coefficients. 170

Table 8.3   Le-Gall 5/3 analysis filter coefficients.   170

Table 8.4   Le-Gall 5/3 synthesis filter coefficients.   171

Table 8.5   Example of sub-bitplane coding order.   179

Table 8.6   Major topics of JPEG 2000 Part 2.   183

Table 10.1   Target bit rates for the JPEG Pleno
holography test set.   214

Table 10.2   Quality metrics according to the type of hologram   217

Table 10.3   Anchor codecs for reference purposes.   218

Table 10.4   Full Bodies point cloud properties.   219

Table 10.5   Upper Bodies point cloud properties..   219

Table 10.6   Small inanimate objects point cloud properties. . . .   220

Table 10.7   Summary of the selected JPEG Pleno l
ight-field images.   227

Table 12.1   LSE marker segment parameter sizes and
values for JPEG-LS preset coding parameters.   245

Table 14.1   Descriptions of the JPSearch core elements.   267

Table 16.1   Entities in the decoding process
(ITU-T-T.88/ISO-IEC-14492, 2018).   284

List of Abbreviations

1D	One-dimensional
2D	Two dimensional
3D	Three dimensional
8iVFB	8i Voxelized Full Bodies
AC	Non-zero frequency coefficient
ACR-HR	Absolute category rating with hidden reference
ADC	Analog to digital conversion
	Analog to digital converter
ATK	Arbitrary transformation kernels
	Autoregressive AR
BAC	Binary arithmetic coding
bps	Bits per sample
CABAC	Context-based adaptive binary arithmetic coding
CBP	Coded block pattern
CDF	Cumulative density function
CI	Common identifier
CIE	International commission on illumination
CTFT	Continuous time Fourier transform
CWT	Continuous wavelet transform
DCT-I	The discrete cosine transform of type I
DCT-II	Disrte cosine transform of type II
DCT-IV	Discrete cosine transform of type IV
DFT	Discrete Fourier transform
DPCM	Differential pulse code modulation
DSIS	Double stimulus impairment scale
DST-VII	discrete sine transform of type VII
DTFT	Discrete time Fourier transform
DTWT	Discrete time wavelet transform
DWT	Discrete wavelet transform
	Discrete Wavelet Transform
	The discrete wavelet transform
EBCOT	Embedded block coding with optimized truncation

En	Box instance number
EOB	End of block
EOI	End of image
EOTF	Electro-optical transfer function
EPFL	École Polytechnique Fédérale de Lausanne
EUSIPCO	European signal processing conference
FB	Filter bank
FCT	Forward core transform
FDCT	Forward discrete cosine transform
FFT	Fast Fourier transform
FIR	Finite impulse response
FLC	Fixed-length code
FOV	Field of view
FPS	frames per second
GIS	Remote sensing and geographic information system
G-PCC	Geometry based point cloud compression
HDR	High dynamic range
	High-Dynamic Range
HEVC	High efficiency video coding
HMD	Head-mounted display
HOEs	Holographic optical elements
HP	High-pass
HSI	Hue, saturation and intensity
HVS	Human visual system
ICASSP	IEEE international conference on acoustics, speech, and signal processing
ICIP	IEEE international conference on image processing
ICME	IEEE international conference on multimedia and expo
ICT	Irreversible color transform
IDCT	Inverse DCT
IDWT	Inverse discrete wavelet transform
IEC	International Electrotechnical Commission
IJG	Independent JPEG Group
ISO	International Organization for Standardization
ISO BMFF	ISO Base Media File Format standard
IVC	Image and video coding group
JBIG	Joint bi-level image experts group
JFIF	JPEG file interchange format
JND	Just noticeable distortion
JPEG	Joint Photographic Experts Group

JPEG XR	JPEG-eXtended Range
JPQF	JPSearch query format
	The JPSearch Query Format
JPS-EM	JPSearch elementary metadata
JPS-MB	JPSearch metadata block
JPTRDL	Translation Rules Declaration Language
KLT	Karhunen–Loève transform
LAN	Local Area Networks
LDR	Low dynamic range
LPS	Less probable symbol
LSB	Least significant bits
LZC	Layered zero coding
MB	Macroblock. *See*
MCU	Minimum coded units
MMSP	IEEE international workshop on multimedia signal processing
MPQF	The MPEG7 Query Format
MPS	Most probable symbol
MSB	Most significant bit
	Most significant bits
MSE	Mean squared errors
MTF	Modulation transfer function
MVUB	Microsoft Voxelized Upper Bodies
nm	nanometer
	Nano-meter
NTSC	National television system committee
PAL	Phase alternation line
PCRD	Post-compression rate-distortion
PCRD-opt	Post-compression rate-distortion optimization
PCS	Picture coding symposium
PCT	Photo core transform
POT	Photo overlap transform
ppi	Pixels-per-inch
PR	Perfect reconstruction
provision	PeRceptually Optimized VIdeo compresSION
PSNR	Peak signal to noise ratio
QoMEX	International conference on quality of multimedia experience
QoS	Quality of service
RCT	Reverisible color transform
RD	Rate-Distortion
RDO	Rate-distortion optimization

RGN	ReGioN of interest marker
RI	Refractive index
RLE	Run-Length encoding
ROI	Region of interest
SBP	Space-bandwidth produc
SECAM	Sequentiel couleur avec mémoire
SLM	Spatial light modulator
SNR	Signal to noise ratio
SPIFF	Still picture interchange file format
SSIM	Structural SIMilarity index
SVCP	Summer school on video compression and processing
SVT	Sveriges Television AB
TCQ	Trellis coded quantization
TMO	Tone mapping operator
	Tone mapping operators
TR	Technical report
TRDL	Translation Rules Declaration Language
TSGD	Two-sided geo-metric distribution
	Two-sided geometric distribution
UMD	Uniformly maximally-decimated
UUID	Universally Unique Identifier
VIF	Visual information fidelity
VIFp	Visual information fidelity in pixel domain
VLC	Variable length coding
VLI	Variable-length integer
V-PCC	Video point cloud coding
WCG	Wide color gamut
WIC	Windows imaging component
YCbCr	Luma, chrominace blue and chrominance red components.
YCoCg	Luma, chrominace green and chrominance orange components.
Z	Packet sequence number

1

Introduction

The field of data compression has seen an explosive growth during the past two decades as the phenomenal advance in the multimedia field. Subbarayappa, S. and Rao, K.R., 2021, February. Video quality evaluation and testing verification of H. 264, HEVC, VVC and EVC video compression standards. In IOP Conference Series: Materials Science and Engineering. Vol. 1045. No. 1, p. 012–028 . IOP Publishing. Kumar, S.N., Bharadwaj, M.V. and Subbarayappa, S., (2021, April). Performance Comparison of Jpeg, Jpeg XT, Jpeg LS, Jpeg 2000, Jpeg XR, HEVC, EVC and VVC for Images. In 6th IEEE International Conference for Convergence in Technology (I2CT), pp. 1–8. IEEE. However, image and video compression is necessary because they occupy most of the bandwidth for their storage or transmission. For example, currently, it is possible to install high-resolution cameras where the captured images and/or video can be sent through local area networks (LAN) thanks to the data compression.

In 1992, the Joint Photographic Experts Group (JPEG) of the International Organization for Standardization (ISO) and the International Electrotechnical Commission (IEC) released the first version of the still image coding standard, informally known as JPEG (ISO/IEC-10918, 1994), based on the discrete cosine transform of type II (DCT-II). The latest version was released in 1994 and consists of seven parts. In 1998, the JPEG LS (ISO/IEC-14495, 1999) was released for lossless and near-lossless compression of continuous-tone still images. JPEG is still the most dominant still image format around and includes two parts.

In 2000, the ISO/IEC released the first edition of the JPEG 2000 (ISO/IEC-15444, 2004) format to improve the compression performance over JPEG and add features such as scalability and editability. JPEG 2000 is based on the discrete wavelet transform (DWT) and defines a set of lossless (bit-preserving) and lossy compression methods for coding bi-level, continuous-tone gray-scale, palletized color, or continuous-tone color digital still images. Until now, JPEG 2000 consists of 14 parts and one more is under development.

1

As technology advances, digital images increase their resolution. The acquisition of higher bit depth (9–16 bits) and their manipulation, largely studied for the last 15 years, begins to attract the interest of the industry arising the need of new formats, which can handle the high dynamic range (HDR) imaging and wide color gamut (WCG) to display a wider range of colors, implemented on various platforms with high multi-generation robustness for both natural and synthetic images and for representing new image modalities, such as texture-plus-depth, light field, point cloud, and holographic imaging.

Since the JPEG 2000, the ISO/IEC continuously updates to accommodate new coding architectures and formats to the growing multimedia demand. This book covers the spectrum from the first JPEG standard to the new standards for still image compression.

1.1 JPEG Series

All image viewers and editors support JPEG (ISO/IEC-10918, 1994) or JPG files. It is the most widely accepted format. Mobile devices also provide support for opening JPEG files. Software such as Apple Preview and Apple Photos can open JPEG files (JPEG, Overview of JPEG, 2018).

JPEG 2000 (ISO/IEC-15444, 2004) is a DWT-based image compression standard. The broadcast market has adopted JPEG 2000 as mezzanine compression in the live production workflows. Other applications of JPEG 2000 are digital cinema; image archives and databases; medical imaging; cultural heritage display; wireless imaging (included as part 11 for efficient transmission of JPEG 2000 imagery over an error-prone wireless network); pre-press industry (pre-press is the process used when digital files are prepared for printing); remote sensing and geographic information system (GIS) for viewing and analysis of multiple layers of spatially related information associated with a geographical location or region; digital photography; scientific and industrial sector, for example, the use of satellite or aerial photography imagery to link to a mapping or GIS system; the Internet included in parts 8, 9, and 10 of this recommendation; surveillance, for example, the use of "region of interest" enhancement to accurate identification of suspects while excluding from analysis, and subsequent public exposure, the innocent bystander; and document imaging applications (JPEG, Overview of JPEG 2000, 2018).

JPEG XS (ISO/IEC-DIS-21122, Under development) is a DWT-based image compression standard of five parts, with very low latency and very low complexity with various degrees of parallelism and it can be implemented on different platforms. JPEG XS is particularly optimized for visually lossless compression for both natural and synthetic images. The typical compression ratios are between 1:2 and 1:6 for both 4:4:4 and 4:2:2 images and image sequences with up to 16-bit component precision. Typical parameterizations address a maximum algorithmic latency between 1 and 32 video lines for a combined encoder–decoder suite. JPEG XS mezzanine codec standard can be applied where, today, uncompressed image data is used; for example, professional video links (3G/6G/12G-SDI), IP transport (SMPTE-ST-2022, 2007), (SMPTE-ST-2110, 2018), real-time video storage, memory buffers, omnidirectional video capture system, head-mounted displays for virtual or augmented reality, and sensor compression for the automotive industry. It also supports resolution up to 8K and features a frame rate from 24 to 120 frames per second (fps) (JPEG, Overview of JPEG XS, 2018).

JPEG XT (ISO/IEC-18477, under publication) is a DCT-based image compression standard. JPEG XT is a nine-part image compression standard primarily designed for continuous-tone photographic content for applications such as compression of images with higher bit depths (9–16 bits), HDR imaging, lossless compression, and representation of alpha channels. Compression ratio of 3:1–8:1 can be achieved. JPEG XT is completely backwards compatible to JPEG (ISO/IEC-14495, 1999) allowing users the choice of working in low or high dynamic range (JPEG, Overview of JPEG XT, 2018).

JPEG XR (ISO/IEC-29199, 2010) is a DCT-based image compression standard and consists of five parts and provides a practical coding technology. JPEG XR is intended for broad use in a very wide range of digital image handling and digital photography applications. Some areas include robust and high fidelity image, HDR workflow, and computationally constrained signal processing environments, such as mobile and embedded applications among other areas (JPEG, Overview of JPEG XR, 2018), (Doufaux, Sullivan, & Ebrahimi, 2009).

JPEG LS (ISO/IEC-14495, 1999) or lossless JPEG is an addition to JPEG standard to enable lossless compression. The standard consists of two parts, the baseline and the extended part. JPEG-LS is intended for low-complexity hardware implementations of very moderate complexity, while, at the same time, providing state-of-the-art lossless compression performance (JPEG, Overview of JPEG-LS, 2018).

JPEG Pleno (ISO/IEC-JTC-1-SC29, 2018) is a standard under development. It aims at providing a standard framework for representing new imaging modalities, such as texture-plus-depth, light field, point cloud, and holographic imaging. Additionally, it also targets to define new tools for improved compression while providing advanced functionality and support for — but not limited to — image manipulation, metadata, image access and interaction, privacy, and security.

The book is divided into 16 chapters and 1 annex. Chapter 1 presents an introduction to the JPEG Series and highlights important applications of the recommendations.

Chapter 2 discusses sampling and quantization theory for 1D and 2D signals. The chapter gives an overview of the human visual system and the most popular color spaces including the color spaces and the YCbCr sample formats used in image and video compression.

Chapter 3 reviews the discrete cosine and sine transforms and their fast algorithms.

Chapter 4 introduces the concepts of entropy coding.

Chapter 5 is an overview of the classical JPEG encoder and decoder and presents some limitations of this recommendation. JPEG is a DCT-based image coding standard. JPEG is a commonly used method of lossy compression for digital images and achieves typically 10:1 compression with little perceptible loss in image quality. However, lossless compression is also possible with JPEG. The first JPEG standard was approved in September 1992 and its benefits make this standard the most popular.

Chapter 6 describes the JPEG-eXtended Range or JPEG XR which is patented by Microsoft under the name HD Photo (formerly Windows Media Photo), and it is the preferred image format for Ecma-388 Open XML Paper Specification documents. JPEG XR uses a type of integer transform employing a lifting scheme. In 2009, this standard was given final approval. The chapter outlines the capabilities of the standard and describes the recommended encoder and decoder parts, the codestream, the profiles, and levels used.

Chapter 7 describes the JPEG XT and specifies backward-compatible extensions of the base JPEG standard. JPEG XT allows mixing of various elements from different profiles in the codestream, allowing extended DCT precision and lossless encoding in all profiles. The standard offers higher bit depths, high dynamic range, lossless and near-lossless compression, and transparency. The critical parts of the standard were published in 2016 and the final part of the standard was published in 2018. This chapter is advocated to describe the encoder and decoder parts and also to explain the lossless coding process and the coding of opacity information.

Chapter 8 outlines the JPEG 2000 standard. JPEG 2000 is a wavelet-based image compression standard and coding system. The standardized filename extension is .jp2 for ISO/IEC 15444-1 conforming files and .jpx for the extended part-2 specifications, published as ISO/IEC 15444-2. The main advantage offered by JPEG 2000 is the significant flexibility of the codestream. A Call for Technical Contributions was issued in March 1997. The "Final Draft International Standard" (FDIS) was ready in August 2000, and, finally, JPEG 2000 became an "International Standard" (IS) in December 2000. Part II was scheduled to become an International Standard in July 2001. This chapter describes the architecture, tiling, and tiles used in JPEG 2000, image pre-processing, the forward inter- and intra-component transform, the quantization process, the subbands partitioning, the entropy coding, and the rate control – also, an overview of coding of the region of interest and error resilience. This chapter gives the applications of JPEG 2000 part 2 and includes the some limitations of this standard.

Chapter 9 describes the JPEG XS standard. JPEG XS is a wavelets-based image compression standard and coding system. JPEG XS defines a compression algorithm with very low latency and very low complexity. JPEG XS can be efficiently implemented on various platforms such as FPGAs, ASICs, CPUs, and GPUs, and it excels with high multi-generation robustness. The typical compression ratios are between 1:2 and 1:6 for both 4:4:4 and 4:2:2 images and image sequences with up to 16-bit component precision. The JPEG XS was introduced in April 2018 by the Belgian company intoPIX. This chapter outlines the architecture of JPEG XS, the multi-component transforms, coefficients pre-quantization and quantization, the rate control mechanism, and the entropy coding process.

Chapter 10 outlines the JPEG Pleno, which is a standard framework for representing plenoptic images, addressing interoperability issues among the plenoptic modalities of light fields, point clouds, and holography. In October 2014, JPEG launched the ground work for JPEG Pleno initiative and the standard is scheduled to be completed by 2020. This chapter overviews the JPEG Pleno, giving a short description of plenoptic representation, and then describes the light field, the point could, and the holographic imaging.

Chapter 11 describes JPEG AIC, which is an experimental still image compression system that combines algorithms from the H.264 and JPEG standards. More specifically, JPEG AIC combines intra-frame block prediction from H.264 with a JPEG-style image coding, followed by context adaptive binary arithmetic coding used in H.264.

Chapter 12 describes the JPEG-LS standard and is aimed at lossless encoding. The main advantage of JPEG-LS over JPEG 2000 is that JPEG-LS is based on a low-complexity Lossless Compression for Images (LOCO-I) algorithm. LOCO-I exploits a concept called context modeling. The JPEG-LS standard also offers a near-lossless mode, in which the reconstructed samples deviate from the original by no more than an amount less than 5. Part 1 of this standard was finalized in 1999 and Part 2 was released in 2003 mainly to introduce extensions such as arithmetic coding. This chapter mainly describes the encoder and decoded architectures of the JPEG-LS standard.

Chapter 13 describes the JPEG XL image coding standard. In 2017, JTC1/SC29/WG1 issued a series of draft calls for proposals on JPEG XL with substantially better compression efficiency (60% improvement) and is expected to follow still image compression performance shown by HEVC HM, Daala, and WebP. The proposals were submitted in September 2018, with current target publication dated in October 2019. This chapter outlines the scope of the standard, use case, requirements, test material and anchors, target rates, objective and subjective image quality testing, and evaluation tools.

Chapter 14 overviews the JPSearch standard and all its parts. JPSearch is a work item of the JPEG and it is an abstract image search and retrieval framework. Interfaces and protocols for data exchange between the components of this architecture are standardized, with minimal restrictions on how these components perform their respective tasks.

Chapter 15 introduces the parts of the JPEG systems standard, which is an overall framework for future and legacy standards to ensure interoperability and functionality exchange between JPEG standards. JPEG systems intend further to specify system layer extensions for JPEG standards to enable generic support for interactivity protocols, high dynamic range (HDR) tools, privacy and security, metadata, 360° images, augmented reality, and 3D.

Chapter 16 describes the JBIG2 image compression standard. JBIG2 is a lossless and lossy image compression standard for bi-level images developed by the Joint Bi-Level Image Experts Group. The bi-level image is segmented into three regions: text, halftone, and generic regions. Each region is coded differently and the coding methodologies are described in this chapter.

2

Digital images

This chapter introduces the concepts of digital images, sampling theory and reconstruction, 2D sampling, and color spaces.

2.1 Introduction

The images compressed with any of the JPEG Series are digital images. Digital images are represented by a two-dimensional (2D) array composed of a fixed number of rows and columns containing finite numbers defined by bits called pixels, picture elements, or samples that also represent the brightness of a given color. The standard digital photo uses an 8-bit range to represent tones. For example, RGB images use 8-bit intensity to represent the color in one of the three channels; in total, RGB images use 24 bits per pixel. Gray-scale images use only 8 bits to represent the range from the darkest gray to the lightest gray tone of one pixel as the images of Figure 2.1.

The number of pixels represents the image size. However, the resolution factor is an extra parameter needed to represent the spatial scale of the image pixels. For example, an image of 3300 × 2550 pixels with a resolution of 300 pixels per inch (ppi) would be a real-world image size of 11" × 8.5". For example, consider the image of Figure 2.1 and suppose that Figure 2.1(a) represents a continuous image; the way we represent this image in a computer is to divide the area into sub-areas called picture elements as shown in Figure 2.1(b). Suppose that the image size is 64 × 96 pixels, and each pixel is represented by a constant. Therefore, we have a piece-wise constant representation of the image. Suppose that we increase the resolution by some means so that we have an image of 128 × 192 pixels as shown in Figure 2.1(c).

It is easy to see that each pixel of the image in Figure 2.1(c) occupies ¼ of the area of each pixel of the image in Figure 2.1(b). We have reduced the pixel area by a factor of 4. In other words, we increased four times the level of details. Naturally, what we observe in Figure 2.1(c) is closer to the original image. The resolution of the image depends on the hardware

7

(a)   (b)   (c)

Figure 2.1 Representation of (a) the original image, (b) the image of 64 × 96 pixels, and (c) image of 128 × 192 pixels.

(distance between pixels) used to acquire it or the algorithm used to find extra details from one level of resolution to the next level of resolution. The study of super-resolution is out of the scope of this book.

2.2 One-Dimensional Sampling

Sampling is the process of collecting numbers that are representative of a signal. These numbers are stored, transmitted, or reproduced in a device. The analog-to-digital conversion (ADC) is an example of use of sampling. Figure 2.2 shows the continuous to discrete conversion of the amplitude of a one-dimensional (1D) signal $x(t)$ taken at a periodic sampling interval T, which is the time separation between two consecutive samples. The signal is not yet completely digital because the values $x[n]$ can still take on any number from a continuous range – that is why we use the terms discrete-time signal here and not digital signal. Afterwards, the amplitudes are retained between sampling intervals and represented using a determined number of bits (bit depth).

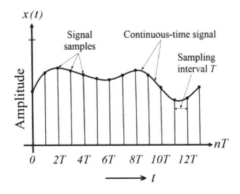

Figure 2.2 Discrete representation of 1D continuous-time signals.

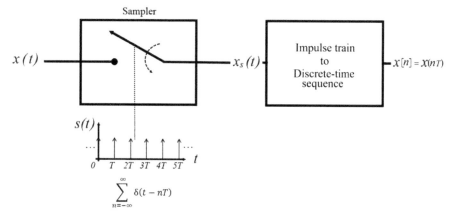

Figure 2.3 Diagram of the ideal continuous to discrete converter (Oppenheim, 2014).

Figure 2.3 represents the block diagram of the continuous to discrete converter. The switch closes every T seconds to let pass a sample of $x(t)$ to $x_s(t)$ and then a conversion from impulse train to discrete time sequence must be carried out to obtain the samples $x[n]$; then $x[n] = x(nT)$ where $-\infty < n < \infty$. In other words, $x[n]$ is the sequence generated by sampling the continuous-time signal $x(t)$ every $t = nT$ seconds. The sampling rate determines the resolution of the discretized signal. In other words, the higher the sampling rate, the higher the resolution. The periodic impulse train is defined as

$$s(t) = \sum_{n=-\infty}^{\infty} \delta(t - nT) = \text{comb}(t; nT) \tag{2.1}$$

where $\delta(t)$ is the Dirac delta function. $s(t)$ is also known as the comb function. The Fourier transform of a comb function is another comb function. The output of the sampler is

$$x_s(t) = x(t) \cdot s(t) \tag{2.2}$$

$$x_s(t) = x(t) \sum_{n=-\infty}^{\infty} \delta(t - nT) \tag{2.3}$$

Let us consider the Fourier transform of $x_s(t)$ which is the convolution of the Fourier transform of $x(t)$ and $s(t)$. The Fourier transform of the comb function of Equation (2.1) is

$$S(j\Omega) = \frac{2\pi}{T} \sum_{k=-\infty}^{\infty} \delta(\Omega - k\Omega_s) = \text{comb}(\Omega) \tag{2.4}$$

Then,

$$X_s(j\Omega) = \frac{1}{2\pi} X(j\Omega) * S(j\Omega) \qquad (2.5)$$

where * is the convolution operator. Therefore,

$$X_s(j\Omega) = \frac{1}{T} \sum_{k=-\infty}^{\infty} X\left(j\left(\Omega - k\Omega_s\right)\right). \qquad (2.6)$$

Note that the sampling frequency is

$$\Omega_s = 2\pi / T$$

Therefore, in Equation (2.6), we see that copies of the input spectrum $X(j\Omega)$ are shifted by integer multiples of Ω_s and then superimposed to produce the periodic Fourier transform of the impulse train of samples. Also, the sampling interval affects the frequency amplitude.

Figure 2.4 shows an example of the (a) spectrum of the input signal or baseband signal with bandwidth Ω_1, commonly referred to as the *Nyquist frequency* (Nyquist, 1928). Figure 2.4(b) is the spectrum of the sampling

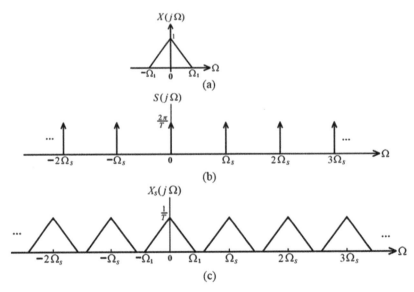

Figure 2.4   Frequency domain of the sampling process. Spectrum of (a) the original signal or baseband signal, (b) the sampling function, and (c) the sampled signal.

frequency Ω_s and Figure 2.4(c) is the spectrum resulting of convolving $X(j\Omega)$ and $S(j\Omega)$ depicted in Equations (2.5) and (2.6). In Figure 2.4, we can see that in order to avoid overlap (aliasing) of the replicas after the sampling process $\Omega_s - \Omega_1 > \Omega_1$. In other words, to avoid aliasing we need

$$\Omega_s > 2\Omega_1. \tag{2.7}$$

Note that

$$f_s > 2f_1$$

where $\Omega_s = 2\pi f_s$ and f_1 is the maximum frequency content in Hertz. The right-hand side of Equation (2.7) is known as the *Nyquist rate*. From Figure 2.4(c), it is easy to see that $x(t)$ can be recovered from $x_s(t)$ using an ideal low-pass filter $H(j\Omega)$ with gain T and cutoff frequency of $\Omega_1 < \Omega_c < (\Omega_s - \Omega_1)$. The filter function is defined by $(\Omega/2\Omega_1)$. For simplicity, the cutoff frequency is the average of Ω_1 and $(\Omega_s - \Omega_1)$. Hence, $\Omega_c = \Omega_s/2 = \pi/T$. Therefore,

$$X_r(j\Omega) = H(j\Omega)X_s(j\Omega). \tag{2.8}$$

Therefore, the recovered spectrum after sampling $X_r(j\Omega)$ will be equal to the spectrum of the original signal $X(j\Omega)$.

Equation (2.2) is the unit impulse train spaced at $t = nT$ and weighted by the samples values $x(nT)$. The multiplication operation yields the following continuous function:

$$x_s(t) = x(t) \cdot s(t) = \sum_{n=-\infty}^{\infty} x(nT)\delta(t - nT). \tag{2.9}$$

Thus, there are two forms of the continuous-time Fourier transform (CTFT), one that expresses the CTFT of $x(t)$ and one that expresses the CTFT of $x_s(t)$. We can write these equations together with the discrete-time Fourier transform (DTFT) and find a relationship to express DTFT in terms of the CTFT of $x_s(t)$

$$X_s(j\Omega) = \sum_{n=-\infty}^{\infty} x(nT)e^{-j\Omega T n} \tag{2.10}$$

$$X(e^{j\omega}) = \sum_{n=-\infty}^{\infty} x[n]e^{-j\omega n} \tag{2.11}$$

Note that $X(e^{j\omega}) = X_s(j\Omega)$ when $\omega = \Omega T$. Hence,

$$X(j\Omega) = X(e^{j})\big|_{\omega=\Omega T} = X(e^{j\Omega T}) \tag{2.12}$$

From Equations (2.12) and (2.6),

$$X(e^{j\Omega T}) = \frac{1}{T} \sum_{k=-\infty}^{\infty} X(j(\Omega - k\Omega_s)) \tag{2.13}$$

or, equivalently,

$$X(e^{j\omega}) = \frac{1}{T} \sum_{k=-\infty}^{\infty} X\left(j\left(\frac{\omega}{T} - \frac{2\pi}{T}k\right)\right). \tag{2.14}$$

If no aliasing occurs, $X(e^{j\omega})$ is a frequency-scaled version of the input spectrum $X_s(j\Omega)$ with frequency scaling $\Omega = \omega/T$. The CTFT $X_s(j\Omega)$ is a periodic function of Ω with a period $\Omega_T = 2\pi/T$. The DTFT $X(e^{j\omega})$ is a periodic function with period 2π. We can state the Nyquist sampling theorem as follows.

Nyquist sampling theorem (Nyquist, 1928): Let $x(t)$ be a band-limited continuous-time signal with $X(j\Omega) = 0$ $|\Omega| \geq \Omega_1$. Then $x(t)$ is uniquely determined by its samples $x[n] = x(nT)$, $n = \cdots -2, -1, 0, 1, 2,\ldots$ if

$$\Omega_s = \frac{2\pi}{T} > 2\Omega_1.$$

We can easily see that $-\frac{1}{2} < \frac{f_1}{f_s} < \frac{1}{2}$ $\rightarrow$ $-\frac{1}{2} < f_d < \frac{1}{2}$. Then, it follows that $-\pi < \Omega_d < \pi$, where Ω_d is the digital frequency. A relation of analog frequency to sampling frequency above ½ results in aliasing.

In signal processing, there are applications with typical bandwidths and sampling frequencies as shown in Table 2.1.

Example 2.1. Determine the Nyquist frequency and Nyquist rate of the following continuous-time signal:

$$x(t) = 1 + \cos(1000\pi t) + \cos(4000\pi t). \tag{2.15}$$

Table 2.1   Typical bandwidths and sampling frequencies.

Applications	f_1 (kHz)	f_s (kHz)
Telephone speech	< 4	8
Ultrasonic	< 100	250
Radar	<100, 100	200,200
Music	< 20	44.1
Biomedical	< 500	1000

The three frequencies in the above equation are 0, 500, and 2000 Hz. The Nyquist frequency is 2000 Hz and the Nyquist frequencies are 2000 and 4000 Hz, respectively (4000π and 8000π).

Example 2.2. Determine the discrete-time sequence $x[n]$ after sampling the continuous-time signal $x(t) = \cos(200\pi t)$ with a sampling period of 1/300s.

$$x[n] = x(nT)$$

$$x(t) = x(nT)\big|_{t=nT}$$

$$x[n] = \cos(200\pi nT) = \cos\left(\frac{200\pi}{300}n\right) = \cos\left(\frac{2\pi}{3}n\right), \quad n = \ldots, -2, -1, 0, 1, 2, \ldots$$

$$x[n] = \{\ldots, -0.5, -0.5, \underset{\uparrow}{1}, -0.5, -0.5, \ldots\}.$$

The frequency of $x[n]$ is $2\pi/3$. The arrow under the number denotes $n = 0$.

Some considerations that are to be observed are that the input signal $x(t)$ may not be band-limited; therefore, an analog low-pass anti-aliasing filter is needed before acquiring the signal. The analog-to-digital converter (ADC) is used to approximate the sampling operation. However, the ADC introduces quantization errors.

2.3 Reconstruction

Figure 2.5 shows the diagram of the ideal reconstruction system. Samples $x[n]$ are sampled to form the impulse train $x_s(t)$ as the product of the periodic unit impulse train of period T and the sequence of samples $x[n]$ as the amplitude of the nth impulse function

$$x_s(t) = \sum_{n=-\infty}^{\infty} x[n]\delta(t - nT) \tag{2.16}$$

$x_s(t)$ is the input to the reconstruction ideal low-pass filter $H_r(j\Omega)$ defined by

$$H_r(j\Omega) = \begin{cases} T & |\Omega_c| \leq \Omega_s / 2 = \pi / T \\ 0 & |\Omega_c| > \Omega_s / 2 = \pi / T \end{cases} . \tag{2.17}$$

The impulse response of the reconstruction filter is the inverse Fourier transform of Equation (2.17) defined by

$$h_r(t) = \frac{\sin(\pi t / T)}{\pi t / T}. \tag{2.18}$$

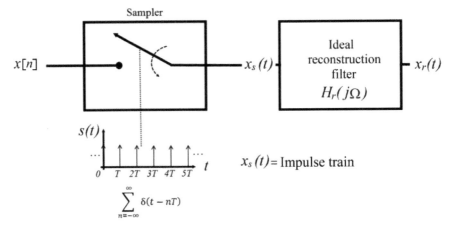

Figure 2.5   Diagram of the ideal reconstruction system (Oppenheim, 2014).

The recovered signal $x_r(t)$ is the convolution of $x_s(t)$ with $h_r(t)$. Then

$$x_r(t) = x_s(t) * h_r(t) = \int_{-\infty}^{\infty} \sum_{n=-\infty}^{\infty} x[n]\delta(\tau - nT) \frac{\sin(\pi\tau/T)}{\pi\tau/T} d\tau$$

$$x_r(t) = \sum_{n=-\infty}^{\infty} x[n] \int_{-\infty}^{\infty} \delta(\tau - nT) \frac{\sin(\pi\tau/T)}{\pi\tau/T} d\tau$$

$$x_r(t) = \sum_{n=-\infty}^{\infty} x[n] \frac{\sin(\pi(t - nT)/T)}{\pi(t - nT)/T}. \qquad (2.19)$$

Equation (2.19) can be verified at $t = nT$. Use the L'Hospital's rule to obtain

$$\frac{\sin(\pi(t - nT)/T)}{\pi(t - nT)/T} = 1.$$

Therefore,

$$x_r(nt) = x[n]$$

when $t \neq nT$

$$\frac{\sin(\pi(t - nT)/T)}{\pi(t - nT)/T} = 0.$$

Equation (2.19) can be expressed in the frequency domain, provided that the samples $x[n]$ were acquired at the Nyquist rate or higher

$$X_r(j\Omega) = \sum_{n=-\infty}^{\infty} x[n]\,H_r(j\Omega)e^{j\Omega T n}\Bigg|_{\omega=\Omega T} = H_r(j\Omega)\sum_{n=-\infty}^{\infty} x[n]e^{j\Omega T n}$$

$$X_r(j\Omega) = H_r(j\Omega)\cdot X(e^{j\Omega T}).$$

The ideal discrete to continuous process is approximated by a discrete to analog converter because the reconstruction (2.19) is impossible because of the infinite summation and the non-causal filtering process.

Example 2.3. Determine two possible continuous-time frequencies Ω_a and Ω_b after sampling $x[n] = \cos\left(\dfrac{\pi n}{8}\right)$ with a sampling frequency of $f_s = 2\text{kHz}$. Implement the MATLAB code for reconstructing f_a in Hertz.

$$x[n] = \cos\left(\frac{\pi n}{8}\right)$$

$$x(nT) = \cos\left(\frac{\pi n}{8}T\right) = \cos\left(\frac{\pi n}{8 f_s}\right) = \cos\left(\frac{\Omega_a}{2000}n\right).$$

Therefore,

$$\frac{\Omega_a}{2000} = \frac{\pi}{8}$$

$$\Omega_a = 250\,\pi \text{ radians/s}$$

$$f_a = 125 \text{ Hz}$$

$$x(t) = \cos\left(250\,\pi t\right).$$

MATLAB code to approximate 4000 amplitudes of $x_r(t)$

```
clear;
N =4000;
n=-200:59;
x=cos(pi.*n./8);
T=1/2000; %sampling interval
xr = zeros(1,N);
for l=1:N
    t=(l-1)*T/100;% sample separation = 0.01T
```

```
      h=sinc((t-n.*T)./T);
      xr(l)=x*h.'; %approximate interpolation
end

time = linspace(0,0.02, N);
[pks,locs] = findpeaks(xr,time);
Freq =strcat(num2str(1/locs(1), 3),' Hertz');

plot(time, xr);
title(Freq);
```

Using the periodicity of the sinusoid

$$\frac{\Omega_b}{2000} = \frac{\pi}{8} + 2\pi$$

$$\Omega_b = 2000\frac{17\pi}{8} = 4250\pi$$

$$f_b = 2125 \text{ Hz.}$$

Example 2.4. Consider the analog sinusoid $x(t) = \cos(2\pi 60t)$ at sampling rates of 240 and 1000 samples per second. The sample spaces are 4.1666 ms for the first sampling rate and 1 ms for the second sampling rate. Program a MATLAB code to show calculations of the results after sampling the analog sinusoid.

```
t = 0:1/2000:.02;
x = cos(2*pi*60*t);
t240 = 0:1/240:.02;
n240 = 0:length(t240)-1;
x240 = cos(2*pi*60/240*n240); % fs = 240 Hz
t1000 = 0:1/1000:.02;
n1000 = 0:length(t1000)-1;
x1000 = cos(2*pi*60/1000*n1000); % fs = 1000 Hz
```

Example 2.5. Consider the four analog sinusoids $x_1(t) = \cos(2\pi 60t + \pi/3)$, $x_2(t) = \cos(2\pi 340t + \pi/3)$, and $x_3(t) = \cos(2\pi 460t + \pi/3)$. Program a MATLAB code to sample the four sinusoids using a sampling frequency of $f_s = 400$Hz and show that the sample values for all three signals are identical.

```
ta = 0:1/4000:2/60; % analog time axis
xa1 = cos(2*pi*60*ta+pi/3);
xa2 = cos(2*pi*340*ta-pi/3);
xa3 = cos(2*pi*460*ta+pi/3);
tn = 0:1/400:2/60; % discrete-time axis as n*Ts
xn1 = cos(2*pi*60*tn+pi/3);
xn2 = cos(2*pi*340*tn-pi/3);
xn3 = cos(2*pi*460*tn+pi/3);
```

2.4 Sampling of Continuous-Time Two-Dimensional Functions

In two-dimensional functions, we have to calculate a function z of two variables, say v and w as

$$z = f(v, w). \tag{2.20}$$

Given the values of v and w, we can obtain a z function; we have three variables, v, w, and z. This is the simplest way to represent a 2D image. For example, the 2D function for $f(v, w) = 20 - 4w$ is shown in Figure 2.6. Note that this function is independent of v. In other words, whatever be the value of v, we will obtain the same values of f.

Columns are constant colors. For example, when $w = 0$, $f(v, w) = 20$, when $w = 1$, $f(v, w) = 20 - 4 = 16$ and so on. In this case, we have taken discrete values of w. However, for a 2D continuous-time function, $v, w \in \mathbb{R}$. Then, an image can be viewed as a 2D function in the form of $f(v, w)$.

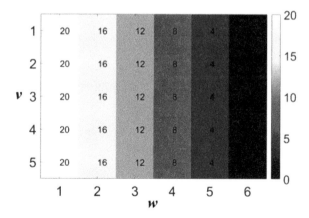

Figure 2.6 Image corresponding to the function $f(v, w) = 20 - 4w$.

Figure 2.7 Image with length W and height H, both in units of length.

Figure 2.7 is an image of length W and height H in units of length. Any point in the image will be identified by the image coordinates v and w being the vertical and horizontal coordinates respectively with limits $0 \leq v \leq H$.

The color value at any point $f(v,w)$ can be represented as

$$f(v,w) = r(v,w) \cdot i(v,w) \tag{2.21}$$

where $r(v,w)$ and $i(v,w)$ are the reflectance of the surface point and the intensity of the light at coordinates v,w. Note that

$$0 \leq r(v,w) \leq 1 \text{ and } 0 \leq i(v,w) \leq \infty. \tag{2.22}$$

Therefore, theoretically, any point, $f(v,w)$ can have a value between 0 and infinity. Nevertheless, the points $f(v,w)$ have a color bounded by a minimum and a maximum value where the maximum value is less than infinity. Therefore, the intensity $i(v,w)$ is bounded by a minimum I_{min} and a I_{max} value. Hence, $I_{min} \leq f(v,w) \leq I_{max}$. Observe that $f(v,w)$, v and w are not discrete variables. Here, we can make use of a uniform grid to discretize the image. We can consider the intensity at each particular point at each of the grid locations. This process is known as sampling. Other sampling grids can be used; however, the uniformly spaced rectangular grid is the most common sampling grid as shown in Figure 2.8. The two-dimensional delta function is defined as (Jain, 1989)

$$\delta(v,w) = \delta(v)\delta(w) \tag{2.23}$$

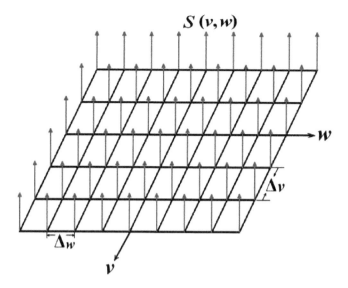

Figure 2.8 Sampling grid.

which satisfies the properties

$$\int_{-\infty}^{\infty}\int_{-\infty}^{\infty} f(v',w') \cdot \delta(v-v',w-w')\, dv'\, dw' = f(v,w) \qquad (2.24)$$

and

$$\lim_{\varepsilon \to 0}\int_{-\varepsilon}^{\varepsilon}\int_{-\varepsilon}^{\varepsilon} \delta(v,w)\, dv\, dw = 1 \qquad (2.25)$$

Therefore, the sampling grid can be defined as

$$s(v,w)=\sum_{m=-\infty}^{\infty}\sum_{n=-\infty}^{\infty} \delta(v-m\Delta v,w-n\Delta w) = comb(v,w;\Delta v,\Delta w) \qquad (2.26)$$

where Δv and Δw are the sampling intervals that need to have no loss of information. Therefore, they depend on the spatial frequency content of the image. Hence, the sampling conditions for no information loss are derived by examining the spectrum of the image in the Fourier domain. After sampling the image, we have

$$f_s(v,w) = f(v,w) \cdot s(v,w) \qquad (2.27)$$

$$f_s(v,w) = \sum_{m=-\infty}^{\infty}\sum_{n=-\infty}^{\infty} f(m\Delta v,n\Delta w) \cdot \delta(v-m\Delta v,w-n\Delta w). \qquad (2.28)$$

The sampling intervals Δv and Δw need to have no loss of information depending on the spatial frequency content of the image. According to Shannon's sampling theorem, in order to preserve the spatial resolution of the original image, the digitizing device must utilize a sampling interval that is no greater than one-half the size of the smallest resolvable feature of the optical image. This is equivalent to acquiring samples at twice the highest spatial frequency contained in the image, a reference point commonly referred to as the Nyquist criterion. Sampling conditions for no information loss can be derived by examining the spectrum of the image by performing the Fourier analysis of $f_s(v,w)$.

Let us assume that the input signal $f(v,w)$ is band-limited, then its Fourier transform is zero outside a bounded region of support (see Figure 2.1) (Jain, 1989), (Oppenheim, 2014).

In Figure 2.9(b), we see that the region of support, in the Fourier domain, of a band-limited 2D signal is defined as

$$F(\Omega_1, \Omega_2) = 0, \quad \text{for } |\Omega_1| > \Omega_{v0}, \quad |\Omega_2| > \Omega_{w0} \tag{2.29}$$

and Ω_{v0}, Ω_{w0} the bandwidths of the signal.

Similar to Equation (2.4), the Fourier transform of a 2D comb function with spacing Δv, Δw is another comb function with spacing $\Omega_{vs} \triangleq 1/\Delta v$, $\Omega_{ws} \triangleq 1/\Delta w$. Therefore, the Fourier transform of Equation (2.26) is

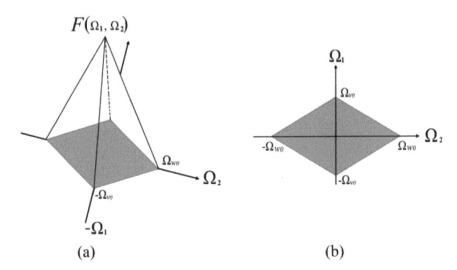

(a)   (b)

Figure 2.9   Fourier transform of (a) a band-limited 2D function and (b) its region of support.

$$COMB(\Omega_1,\Omega_2) = \Im\{comb(v,w;\Delta v,\Delta w)\} = \Omega_{vs} \cdot \Omega_{ws} \cdot comb(\Omega_v,\Omega_w;\frac{1}{\Delta v},\frac{1}{\Delta w}).$$

(2.30)

Therefore, the Fourier transform of Equation (2.27) is the convolution of

$$F_s(\Omega_1,\Omega_2) = F(\Omega_1,\Omega_2) * COMB(\Omega_1,\Omega_2)$$ (2.31)

$$F_s(\Omega_1,\Omega_2) = \int_{-\infty}^{\infty}\int_{-\infty}^{\infty} f(v,w) \cdot \left(\sum_{k=-\infty}^{\infty}\sum_{l=-\infty}^{\infty} \frac{1}{\Delta v}\frac{1}{\Delta w}\exp\left[j2\pi\frac{kv}{\Delta v}\right]\exp\left[j2\pi\frac{lw}{\Delta w}\right]\right)$$
$$\cdot \exp\left[-j2\pi\Omega_1 v\right]\exp\left[-j2\pi\Omega_2 w\right] dv\,dw$$

(2.32)

$$F_s(\Omega_1,\Omega_2) = \frac{1}{\Delta v}\frac{1}{\Delta w}\sum_{k=-\infty}^{\infty}\sum_{l=-\infty}^{\infty}\int_{-\infty}^{\infty}\int_{-\infty}^{\infty} f(v,w)\exp\left[j2\pi v\left(\Omega_1 - \frac{k}{\Delta v}\right)\right]$$
$$\cdot \exp\left[j2\pi w\left(\Omega_2 - \frac{l}{\Delta w}\right)\right]dv\,dw$$

$$F_s(\Omega_1,\Omega_2) = \frac{1}{\Delta v}\frac{1}{\Delta w}\sum_{k=-\infty}^{\infty}\sum_{l=-\infty}^{\infty} F\left(\Omega_1 - \frac{k}{\Delta v},\Omega_2 - \frac{l}{\Delta w}\right)$$

$$F_s(\Omega_1,\Omega_2) = \Omega_{vs}\,\Omega_{ws}\sum_{k=-\infty}^{\infty}\sum_{l=-\infty}^{\infty} F\left(\Omega_1 - k\Omega_{vs},\Omega_2 - l\Omega_{ws}\right).$$ (2.33)

Note, in Equation (2.33), the scaled and periodic replications of the Fourier transform of the input signal $F(\Omega_1,\Omega_2)$, on a frequency grid, with separations of multiples of Ω_{vs},Ω_{ws} (see Figure 2.10).

2.5 Image Reconstruction from Its Samples

From the sampled image spectrum in Figure 2.10, we can see that it is possible to recover the image spectrum of Figure 2.9 if the sampling frequencies are greater than twice the bandwidths.

$$\Omega_{vs} > 2\Omega_{v0}, \quad \Omega_{ws} > 2\Omega_{w0}.$$ (2.34)

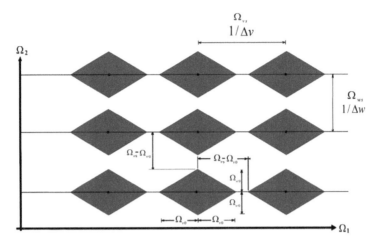

Figure 2.10   Replicas of the sampled image spectrum with no aliasing.

Therefore, $F(\Omega_1, \Omega_2)$ can be recovered by low-pass filtering the sampled image. The region of support of the filter is the rectangle centered at the origin and defined by

$$\left[-\frac{1}{2}\Omega_{vs}, \frac{1}{2}\Omega_{vs}\right] \times \left[-\frac{1}{2}\Omega_{ws}, \frac{1}{2}\Omega_{ws}\right]. \tag{2.35}$$

The impulse response of the filter is

$$h(v, w) = \operatorname{sinc}\left(v\Omega_{vs}\right)\operatorname{sinc}\left(w\Omega_{ws}\right) \tag{2.36}$$

$$h(v, w) = \frac{\sin\left(\pi v\, \Omega_{vs}\right)}{\pi v \Omega_{vs}} \cdot \frac{\sin\left(\pi w\, \Omega_{ws}\right)}{\pi w\, \Omega_{ws}}. \tag{2.37}$$

The reconstructed image is the convolution of the sampled image with the impulse response of the ideal filter.

$$\hat{f}(v, w) = f_s(v, w) * h(v, w) \tag{2.38}$$

$$\hat{f}(v, w) = \sum_{m=-\infty}^{\infty}\sum_{n=-\infty}^{\infty} f_s(m\Delta v, n\Delta w) h(v - m\Delta v, w - n\Delta w) \tag{2.39}$$

$$\hat{f}(v, w) = \sum_{m=-\infty}^{\infty}\sum_{n=-\infty}^{\infty} f_s(m\Delta v, n\Delta w)\frac{\sin\left(\pi\left(v\, \Omega_{vs} - m\right)\right)}{\pi\left(v\, \Omega_{vs} - m\right)} \cdot \frac{\sin\left(\pi\left(w\, \Omega_{ws} - n\right)\right)}{\pi\left(w\, \Omega_{ws} - n\right)}. \tag{2.40}$$

Since the sinc function has infinite extent, it is impossible in practice to implement the ideal LPF. Therefore, it is impossible to reconstruct the exact image from its samples without error if we sample it at the Nyquist rates. Nevertheless, a practical solution is to sample the image at higher spatial frequencies and implement a real low-pass filter as close as possible to the ideal low-pass filter.

Example 2.6. One way to classify signals is by the dimension of the domain of the function, i.e., how many arguments the function has.
- A one-dimensional signal is a function of a single variable, e.g., time or elevation above sea level. In this case, the range is a subset of $\mathbb{R} = (-\infty, \infty)$, the set of all real numbers.
- An M-dimensional signal is a function of M independent variables. In this case, the range is a subset of $\mathbb{R}^M = (-\infty, \infty)^M$, the set of all M-tuples of real numbers.

Example 2.7. A sequence of "black and white" (gray-scale) TV pictures $I(x,y,t)$ is a scalar-valued function of three arguments: spatial coordinates x and y, and time index t; so it is a 3D signal. We focus primarily on two-dimensional signals, i.e., images, generally considering the independent variables to be spatial position (x, y).

Example 2.8. Another way to classify signals is by the dimension of the range of the function, i.e., the dimension of the space of values the function can take scalar for single-channel signals or multichannel signals; i.e., a gray-scale TV picture is scalar valued, whereas a color TV picture can be described as a three-channel signal, where the components of the signal represent red, green, and blue (RGB). Here, $g(x,y) = \left[g_R(x,y), g_G(x,y), g_B(x,y) \right]$. We will consider both scalar (gray-scale) images and multichannel (e.g., color) images.

2.6 Image Quantization

The term quantization refers to the technique to make the range of a signal finite. A quantizer maps a continuous variable f taking values in the range $\{t_1, \ldots, t_{L+1}\}$ or $f \in [t_1, t_{L+1}]$ into a discrete variable f' taking values in the rage $\{r_1, \ldots, r_L\}$ or $f' \in [r_1, r_L]$ (Jain, 1989); these values are known as reconstruction values. After sampling, the image $f(v,w)$ must be digitized both spatially and in amplitude. The sampling rate determines spatial resolution and the quantization level determines the number of gray levels or colors in the digital image. Unlike sampling (above the Nyquist rate), quantization is, in general, a non-reversible operation; in other words, it is a lossy operation (Oppenheim, 2014), (Jain, 1989). The number of quantization levels can reduce or increase the amount of data needed to represent each sample and its choice establishes the quality of the quantized image. The mapping rule is the set of *decision levels* $\{t_k : k = 1, \ldots, L+1\}$ with right open intervals $[t_k, t_{k+1}]$ of *f* in the *k*th reconstruction level r_k.

In other words, if the amplitude of *f* lies in the interval $[t_k, t_{k+1}]$ it is mapped to r_k. Note that this mapping operation is many to one. Figure 2.11 shows an example of the mapping function; this function is known as the midrise quantizer. In the midrise quantizer, the origin lies in the middle of a raising part of the staircase-like graph; zero is not included in the reconstruction levels and the number of quantization levels (*L*) is even. The input to the quantizer is the function *f* with a probability distribution and the outputs are the quantized amplitudes.

The design of a quantizer includes the following information:
- the input brightness range $f \in [t_1, t_{L+1}]$;
- the number of bits (*B*) of the quantizer to calculate the number of reconstruction levels $L = 2^B$;
- an expression for the decision levels;
- an expression for the reconstruction levels.

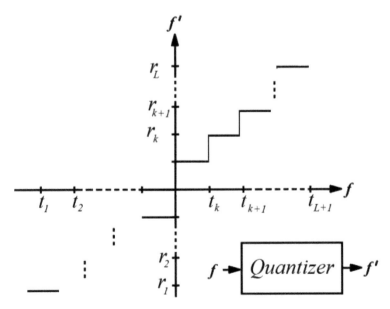

Figure 2.11   Midrise quantizer.

2.6.1 The uniform quantizer

The uniform quantizer assumes that gray-scale or color scale of the input follows a continuous distribution between $[t_1, t_{L+1}]$. Therefore, the intervals $[t_k, t_{k+1}]$ are equal. The uniform distribution is defined as

$$p_f(f) = \begin{cases} \dfrac{1}{t_{L+1} - t_1}, & t_1 \leq f \leq t_{L+1} \\ 0, & \text{otherwise} \end{cases} \tag{2.41}$$

In Figure 2.11, we see that each interval is defined by a constant value. For example, the interval $[t_k, t_{k+1}]$ has an optimum reconstruction value (Jain, 1989) equal to

$$r_k = \frac{t_{k+1} + t_k}{2} \tag{2.42}$$

and also

$$t_k = \frac{t_{k+1} + t_{k-1}}{2}. \tag{2.43}$$

From Equation (2.43), we determine the quantization step

$$t_k - t_{k-1} = t_{k+1} - t_k = \text{constant} \triangleq q.$$

Each constant has an equal space of length

$$q = \frac{t_{L+1} - t_1}{L}.$$

Therefore, decision and reconstruction levels are equally spaced.

$$t_k = t_{k-1} + q, \qquad r_k = t_k + \frac{q}{2}.$$

Example 2.4. Let the signal f be uniformly distributed in the interval $[t_1, t_{L+1}] = [0, 256]$. Design the decision and reconstruction table for a 2-bit quantizers. Show the transfer function. Check your results with Table 2.2.

$$B = 2, \ L = 4, \qquad q = \frac{256 - 0}{4} = 64, \quad t_k = t_{k-1} + 64, \quad r_k = t_k + 32.$$

Figure 2.12 shows the decision and reconstruction levels for $B = 2$ of an optimal uniform quantizer.

Table 2.2   Decision and reconstruction levels for $B = 2$.

Decision levels	Reconstruction levels
$t_1 = 0$	$r_1 = 32$
$t_2 = 64$	$r_2 = 96$
$t_3 = 128$	$r_3 = 160$
$t_4 = 192$	$r_4 = 224$
$t_5 = 256$	224

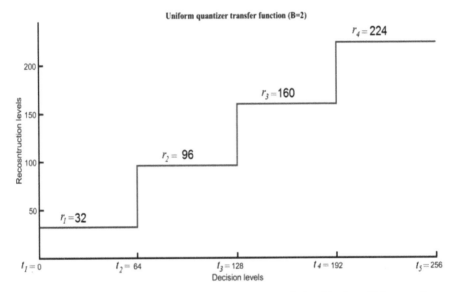

Figure 2.12   Uniform quantizer transfer function for $B = 2$, $L = 4$ in the midrise quantizer.

2.6.2 Quantization error

Quantization error or distortion is the MSE between the non-quantized (f) and the quantized signal (f')

$$E\left[(f - f')^2\right] = \frac{1}{M \times N} \sum_{v=0}^{M-1} \sum_{w=0}^{N-1} \left[(f(v,w) - f'(v,w))^2\right] \quad (2.44)$$

$$= \sum_{k=1}^{L} \int_{tk}^{tk+1} (f - r_k)^2 \; p_f(f) \; df. \quad (2.45)$$

In order to minimize Equation (2.45) over a set r_k, it is necessary to choose the reconstruction value that minimizes the corresponding integral. For an

L-level uniform quantizer, when the random variable is distributed in the interval $[-A,A]$, and quantization steps of size $q = \dfrac{2A}{L}$, the distortion is

$$\text{MSE} = \int_{-q/2}^{q/2} f^2\, p_f(f)df = \frac{1}{q}\int_{-q/2}^{q/2} f^2\, df = \frac{q^2}{12} \tag{2.46}$$

and the mean

$$E[f] = \frac{1}{2A}\int_{-A}^{A} f^2\, df = \frac{A^2}{3}. \tag{2.47}$$

Finally, for uniform optimal quantizers, we find the optimal value q that minimizes the distortion and the optimal reconstruction values within each region.

Example 2.5. Let the signal f be uniformly distributed in the interval $[t_1, t_{L+1}] = [0,\ 256]$. Design the decision and reconstruction levels table for a 3-bit uniform optimal quantizers. Check your results with Table 2.3

$$B = 3,\quad L = 8,\quad q = \frac{256 - 0}{8} = 32,\quad t_k = t_{k-1} + 32,\quad r_k = t_k + 16.$$

Figure 2.13 shows an example of an image quantized using the quantizer of Examples 2.4 and 2.5. The resulting mean squared errors (MSE) are 317.13 and 92.79, respectively.

The bit rate was reduced from 8 bits per sample (bps) to 2 and 3 bps, respectively. Observe that, as we decrease the number of gray levels, some false colors or edges start appearing on the image. These lines are known as contouring. If an image has more detail, the effect of contouring starts appearing later under quantization, as compared to an image with less detail. Therefore, effect of contouring not only depends on the decreasing of gray-level resolution but also on the image detail.

Table 2.3 Decision and reconstruction levels for $B = 3$.

Decision levels	Reconstruction levels
$t_1 = 0$	$r_1 = 16$
$t_2 = 32$	$r_2 = 48$
$t_3 = 64$	$r_3 = 80$
$t_4 = 96$	$r_4 = 112$
$t_5 = 128$	$r_5 = 144$
$t_6 = 160$	$r_6 = 176$
$t_7 = 192$	$r_7 = 208$
$t_8 = 224$	$r_8 = 240$
$t_9 = 256$	240

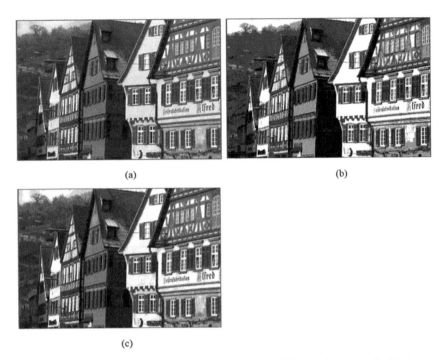

Figure 2.13   Images (a) original, (b) quantized using 2 bits or 4 gray levels, MSE = 337.13, and (c) quantized using 3 bits or 8 gray levels, MSE = 92.79.

Figure 2.14 shows the histogram of frequencies of the gray-scale of the original image in Figure 2.13(a). The vertical lines denote the boundaries of the decision levels and the numbers inside each bin are the reconstruction levels of a 2-bit uniform optimal quantizer. Note that each bin contains a different number of tones; therefore, the image cannot be treated as an uniformly distributed signal. Figure 2.15 shows the reconstruction values of a 3-bit uniform optimal quantizer.

2.6.3 The Optimal Quantizer (Lloyd–Max Quantizer)

The primary goal of the quantizer design is to select the decision and reconstruction levels that yield the minimum possible average distortion, for a fixed number of levels L or a fixed resolution. An optimal quantizer must be designed to exploit the well-defined structures of the input probability distributions to improve the coding gain. For example, in case of uniform quantizers, it assumes that the input values have the same probability of

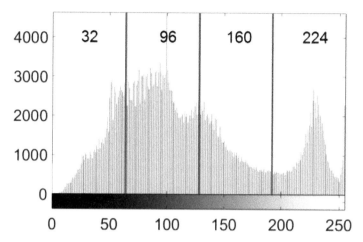

Figure 2.14 Histogram of frequencies of the gray-scale of the original image in Figure 2.13(a). The vertical lines denote the boundaries of the decision levels and the numbers inside of each bin of the reconstruction levels of a 2-bit uniform optimal quantizer.

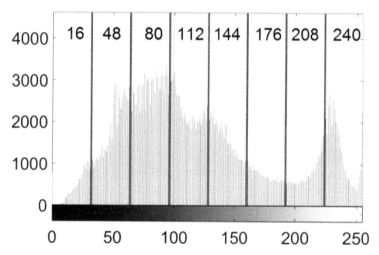

Figure 2.15 Histogram of frequencies of the gray-scale of the original image in Figure 2.13(a). The vertical lines denote the boundaries of the decision levels and the numbers inside of each bin of the reconstruction levels of a 3-bit uniform optimal quantizer.

occurrence. However, in the limit, images follow a Gaussian distribution and a natural image in the DCT domain follows a Laplacian distribution. If we assume zero mean signals, most of the time, the random variable is around zero rather than in the tail of the probability distribution increasing the quantization error according to Equation (2.45). Therefore, a uniform quantizer is neither suitable to quantize images in their natural domain nor in their transform domain. The Lloyd-Max quantizer is known as the optimum mean square quantizer because it minimizes the mean square error for a given number of levels with an input distribution $p_f(f)$. The necessary conditions for minimization of Equation (2.45) are obtained by differentiating with respect to r_k and t_k and setting the results to zero

$$\frac{\partial}{\partial t_k} E\left[(f - f')^2\right] = \left[(t_k - r_{k-1})^2 - (t_k - r_k)^2\right] p_f(t_k) = 0$$

$$\frac{\partial}{\partial r_k} E\left[(f - f')^2\right] = 2\int_{tk}^{tk+1} (f - r_k) p_f(t_k) du = 0, \quad 1 \le k \le L.$$

Assuming $t_{k-1} \le t_k$, then

$$t_k = \frac{(r_k + r_{k-1})}{2}. \tag{2.48}$$

Boundaries of the decision regions are the midpoint of the quantization values

$$r_k = \frac{\int_{tk}^{tk+1} f\, p_f(f)\, df}{\int_{tk}^{tk+1} p_f(f)\, df} = E\left[f \mid f \text{ in the } k\text{th interval } [t_k, t_{k+1})\right]. \tag{2.49}$$

The optimum reconstruction levels lie halfway between the optimum decision levels. The decision levels lie at the center of mass of the probability density in between reconstruction levels. Clearly, Equation (2.48) depends on Equation (2.49) and vice versa. The following algorithm can be used for finding the optimal quantizer:

- Start with arbitrary reconstruction levels (i.e., uniform q and $r_1 < r_2 \cdots < r_L$).
- Find optimal decision values.

$$\text{Set } t_k = \frac{(r_k + r_{k-1})}{2}, \quad k = 1, \ldots, L+1.$$

- Calculate new reconstruction levels,

$$r_k = \frac{\int_{tk}^{tk+1} f\, p_f(f)\, df}{\int_{tk}^{tk+1} p_f(f)\, df}, \quad k = 1, \ldots, L.$$

- Repeat 2 and 3 until distortion between previous and current iterations is negligible.

Two probability densities commonly used are the Gaussian and the Laplacian distributions that are defined as

$$\textbf{Gaussian: } p_f(f) = \frac{1}{\sqrt{2\pi\sigma^2}} \exp\left(\frac{-(f-\mu)^2}{2\sigma^2}\right)$$

where μ and σ^2 are the mean and variance of f.

$$\textbf{Laplacian: } p_f(f) = \frac{\alpha}{2}\exp\left(-\alpha|f-\mu|\right).$$

The variance of the Laplacian distribution is $\sigma^2 = 2/\alpha$.

2.6.4 The Mid-Tread Quantizer

In the mid-tread quantizer, the origin lies in the middle of a tread of the staircase-like graph. The number of quantization levels (L) in this type is odd.
    Figure 2.16 shows the (a) midrise and (b) the midtread quantizers.

Example 2.6: The signal input to a 2-bit uniform quantizer fluctuates in the interval $[-1, 1]$:
  a) A 2-bit uniform midrise quantizer for the input intervals, $[-1, -0.5)$, $[-0.5, 0.0)$, $[0.0, 0.5)$, $[0.5, 1]$, produces the output $\{11, 10, 00, 01\}$ with reconstruction values $\{-3/4, -1/4, 1/4, 3/4\}$.

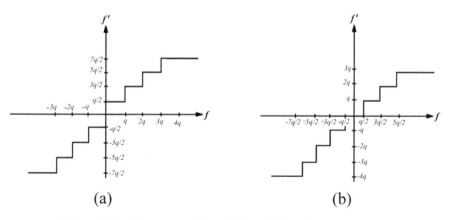

(a)   (b)

Figure 2.16   Comparison of (a) midrise and (b) midtread quantizers.

b) A 2-bit uniform mid-tread quantizer, for the input intervals [−1, −1/3), [−1/3, 1/3), [1/3, 1], produces the output {11, 00/10, 01} with reconstruction values {−2/3, 0, 2/3}.

The subinterval size of the mid-tread quantizer is larger than that of the midrise quantizer. If the input signal is uniformly distributed, the average quantization error is larger than the quantization error of the midrise quantizer. In coding system design, a quantizer that represents signals with zero amplitude performs better than the one that cannot because quantizers are applied after the input signal has been decorrelated (transformed). Hence, the input to the quantizer receives values with magnitudes that can be easily represented with zero.

2.7 The Human Visual System

The human visual system (HVS) is part of the human nervous system and refers to the human's visual perception that allows us to see. The HVS has been modeled and included in computer systems. For example, in a lossy coding scheme, the encoder removes the information not seen by the human eye, yielding more compression without sacrificing the visual quality of the reconstructed signal.

The HVS was first incorporated by (Mannos & Sakrison, 1974) in image coding. Since then, other works have been carried out (Nill, 1985), (Dalay, 1987), (Ngan, Leong, & Singh, 1989), (Long- Wen, Ching- Yang, & Shiuh-Ming, 1999), (Xinfeng *et al.* 2017) engulfing the same topic for different applications (Kumar & Naresh, 2016), (Chitprasert & Rao, 1990). For instance, the quantization table recommended by JPEG standard (Wallace, 1992) tends to preserve low-frequency information and discard high-frequency details because HVS is less sensitive to the information loss in high-frequency bands. Assuming that the HVS is isotropic, Mannos and Sakrison (1974) modeled the HVS as a nonlinear point transformation followed by the modulation transfer function (MTF) given by

$$a(b + cF)\exp\left(-c\,F^{d}\right) \tag{2.50}$$

where F is the radial frequency of the visual angle subtended in cycles/degree; a, b, c, and d are constants. The sensitivity of the eye to changes in luminance intensity is shown in Figure 2.17. The plot shows the relative response of the eye to threshold intensity changes at different angular frequencies.

Above 0.5 cycles/degree, contrast sensitivity is greater to luminance than to chromatic gratings. Below 0.5 cycles/degrees, contrast sensitivity is

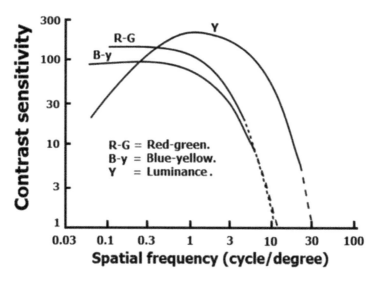

Figure 2.17 Contrast sensitivity functions of luminance signals, and red-green and blue-yellow opponent chromatic signals (McKeefry, Murray, & Kulikowski, 2001) (dash lines are extrapolated experimental data).

greater to the chromatic gratings, consisting of two monochromatic gratings added in antiphase, than to either monochromatic grating alone. The eye is unable to detect any point above the curve. Below the curve, the eye can see the variations in intensity. For horizontal and vertical patterns, the frequency response is similar. For diagonal patterns, it has been shown that the response is significantly reduced. HVS has been used in some compression systems, JPEG among them (Hass, 2018), to derive perceptual quantization tables for the DCT coefficients.

2.7.1 HVS-Based Quantization

The work by Long-Wen, Ching-Yang, and Shiuh-Ming (1999) is based on a procedure to design a quantization table based on the human visual system model for the baseline JPEG coder and demonstrated that HVS-based quantization table can achieve better performance in rate-distortion sense than the JPEG default quantization table. They used Equation (2.50) with $a = 2.2$, $b = 0.192$, $c = 0.114$, $d = 1.1$, and the MTF as reported by Daly (1987) given by

$$H(v,w) = \begin{cases} 2.2\left[0.192 + 0.114\,\tilde{F}(v,w)\right]\exp\left(-\left[0.114\tilde{F}(v,w)\right]^{1.1}\right) & \text{if } \tilde{F}(v,w) > F_{max} \\ 1.0 & \text{otherwise} \end{cases}$$

$$(2.51)$$

where $\tilde{F}(v,w)$ is the radial spatial frequency in cycles/degree and $F_{max} = 8$ cycles/degree at which the exponential peak is. The horizontal and vertical discrete frequencies in the DCT domain, for a symmetric grid, are periodic and given in terms of the dot pitch Δ of the output device and N is the number of frequencies by

$$F(v) = \frac{v-1}{\Delta - 2N}, \qquad v = 1,2,\ldots,N$$

$$F(w) = \frac{w-1}{\Delta - 2N}, \qquad w = 1,2,\ldots,N.$$

(2.52)

Converting these to radial frequencies and scaling the result to cycles/visual degree for a viewing distance, dis, in millimeters gives

$$F(u,v) = \frac{\pi}{180 \sin^{-1}\left(\frac{1}{\sqrt{1+\text{dis}^2}}\right)} \sqrt{F(v)^2 + F(w)^2}, \qquad u,v = 1,2,\ldots,N.$$

(2.53)

Finally, to account for variations in visual MTF as a function of viewing angle, θ, these frequencies are normalized by an angular-dependent function, $S\big(\theta(v,w)\big)$, such that

$$\tilde{F}(v,w) = \frac{F(v,w)}{S\big(\theta(v,w)\big)}.$$

(2.54)

$S\big(\theta(v,w)\big)$ is given by Daly (1987) and Sullivan, Ray, and Miller (1991)

$$S\big(\theta(v,w)\big) = \frac{1-x}{2}\cos\big(4\theta(v,w)\big) + \frac{1+x}{2}.$$

(2.55)

Note that x is a symmetric parameter. If x decreases, $S\big(\theta(v,w)\big)$ decreases near $45°$, $\tilde{F}(v,w)$ increases, and $H(v,w)$ decreases, making the visual MTF more low-pass at those angles. Note that

$$\theta(v,w) = a\tan\left(\frac{F(w)}{F(v)}\right).$$

(2.56)

The human visual frequency-weighting matrix $H(v,w)$ indicates the perceptual importance of the DCT coefficients. After multiplying the 64 DCT coefficients by this weighting matrix, we can summarize that the weighted DCT coefficients contribute the same perceptual importance to human

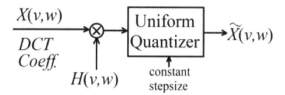

Figure 2.18 Human visual frequency weighting of DCT coefficients.

observers. Therefore, a uniform quantizer with constant step size is employed to quantize the weighted DCT coefficients, as shown in Figure 2.18.

From Figure 2.18, we have

$$\tilde{X}(v,w) = \text{round}\left\lfloor \frac{X(v,w) \cdot H(v,w)}{q} \right\rfloor, \tag{2.57}$$

where q is the quantizer step size. We can rewrite Equation (2.57) as

$$\tilde{X}(v,w) = \text{round}\left\lfloor \frac{X(v,w) \cdot H(v,w)}{Q(v,w)} \right\rfloor, \tag{2.58}$$

where

$$Q(v,w) = \frac{q}{H(v,w)}. \tag{2.59}$$

Then, the HVS-based quantization table Q_{HS} can be derived by

$$Q_{HS} = \text{Round}\left[Q(v,w)\right] \text{ for } 0 \leq v,w < N \tag{2.60}$$

where N is the block size. In JPEG, $N=8$, for $q = 16$, the output of the HVS-based quantizer, for the 63 AC luminance coefficients with non-zero frequencies is

$$\tilde{X}(v,w) = \text{round}\left\lfloor \frac{X(v,w) \times 16}{Q_{HS}(v,w) \times qs} \right\rfloor, \tag{2.61}$$

qs is a scale factor, as defined in JPEG extensions (ISO/IEC-10918, 1994), to control the bit rate. The human visual frequency weighting matrix $H(v,w)$ for a dot pitch of about 0.25 mm 100 dpi, assuming the display aspect ratio 1:1 with dimension of 128 mm × 128 mm to display a 512 × 512 pixel image is shown in Table 2.4.

Table 2.4 The human visual frequency-weighting matrix $H(v,w)$ (Long- Wen, Ching-Yang, & Shiuh-Ming, 1999).

					w			
	1.0000	1.0000	1.0000	1.0000	0.9599	0.8746	0.7684	0.6571
	1.0000	1.0000	1.0000	1.0000	0.9283	0.8404	0.7371	0.6306
	1.0000	1.0000	0.9571	0.8898	0.8192	0.7371	0.6471	0.5558
v	1.0000	1.0000	0.8898	0.7617	0.6669	0.5912	0.5196	0.4495
	0.9599	0.9283	0.8192	0.6669	0.5419	0.4564	0.3930	0.3393
	0.8746	0.8404	0.7371	0.5912	0.4564	0.3598	0.2948	0.2480
	0.7684	0.7371	0.6471	0.5196	0.3930	0.2948	0.2278	0.1828
	0.6571	0.6306	0.5558	0.4495	0.3393	0.2480	0.1828	0.1391

Table 2.5 HVS-based luminance quantization table $Q_{HS}(v,w)$ (Long-Wen, Ching-Yang, & Shiuh-Ming, 1999).

					w			
	16	16	16	16	17	18	21	24
	16	16	16	16	17	19	22	25
	16	16	17	18	20	22	25	29
v	16	16	18	21	24	27	31	36
	17	17	20	24	30	35	41	47
	18	19	22	27	35	44	54	65
	21	22	25	31	41	54	70	88
	24	25	29	36	47	65	88	115

The resulting HVS-based luminance quantization table $Q_{HS}(v,w)$ is shown in Table 2.5. The chrominance quantization tables can be generated by the same method.

While we can distinguish about two dozens of intensity or gray shades, we can perceive thousands of colors and color shades. Therefore, color is a very important topic in all areas of image processing. In the next section, we will study the different representations of color, which is important to represent more visual information of images.

2.8 Color Space

A color space is a mathematical model that describes the range of colors. It is the range (spectrum) of colors that can be represented by a digital device, a printer, or an image. Each color is ordered in tuples of numbers (3 or 4 numbers). Some common color spaces are RGB, YCbCr, YUV, CMY, HSV, YPbPr, and HIS. For example, the RGB (R = Red, G = Green, B = Blue) color space uses three components red, green, and blue, to build a color model where each pixel of an image assigns a range from 0 to 255 intensity

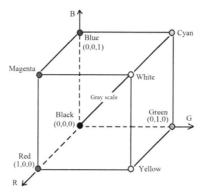

Figure 2.19 Cartesian coordinate system of RGB color space.

values to each component for a total of 16,777,216 colors on the screen by different mixing ratios.

A color model can be represented as a coordinate system within which any specified color will be represented by a single point. For example, the RGB model is based on the Cartesian coordinate system where red, green, and blue components are placed along one coordinate axis as shown in Figure 2.19. For example, red is in the direction of (1,0,0), green in the direction of vector (0,1,0), and blue in the direction (0,0,1) building a 3D Cartesian coordinate system.

Cyan, magenta, and yellow are at the other three corners of the Cartesian coordinate system. Note that the points are normalized within the range 0–1. The origin of the model is (0,0,0) and represents black. The farthest vertex is (1,1,1) and represents white. The joint straight line connecting the points (0,0,0) and (1,1,1) represents the gray-scale or intensity axis.

All the colors of the visible light or the visible spectrum, color spectrum occupy a very narrow spectrum in the total electromagnetic band of frequencies or band of spectrum and the visible spectrum, and the wavelength normally varies from 400 to 700 nm. We normally take three color components (red, green, and blue) as the primary color components because, in our eye, there are three types of cone cells responsible for the red light, green light, and blue light sensation, respectively. The lights are mixed together in different proportions in an appropriate way so that we can have the sensation of different colors. By mixing these three primary colors in different proportion, we can generate almost all the colors in the visible spectrum.

The RGB color space is mostly useful for the display purpose. However, for color printing, the model used is the CMY model or cyan, magenta, and yellow model and the CMY model can be generated from the RGB model.

The CMY is equal to (1, 1, 1) minus RGB. The color model CMY does not produce a pure black but a muddy black. Hence, the component black must be specified to construct the model CMYK with four components.

The RGB color model is oriented toward the color display or color monitor. The CMY and CMYK models are oriented toward color printers. However, we do not really think of how much of red, how much of green, and how much of blue is contained within that particular color. What we really think of a color is what the prominent color is in a specified color.

Hue tells us the prominent primary color or spectrum color in a particular specified color. Saturation indicates how much a pure spectrum color is diluted by mixing white color to it. Therefore, if you mix white colors to appear spectrum color in different amounts, what we get is different shades of that particular spectrum color. The intensity is the chromatic motion of brightness of black and white image. These three features compose another color space called hue, saturation, and intensity (HSI). This model also decouples intensity (luminance) information from color information (hue and saturation). Therefore, the conversion from RGB to HSI model is important.

In order to do this, we reorient the RGB cube of Figure 2.19 in such a way that the black point or the origin in the RGB cube is kept at the bottom and the white comes directly above it as shown in Figure 2.20. Observe the vertical line joining the black and white points is the intensity axis and does not contain any color component.

Suppose that we want to convert any RGB color point P in the reoriented RGB cube. First, we draw a vector into starting from the black vertex. Then,

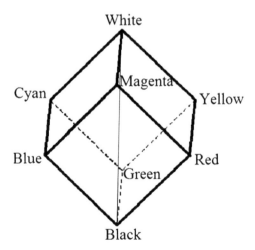

Figure 2.20   RGB–HSI relation.

we project the vector into the intensity axis and the length of the projection is the intensity component associated with the RGB color point. Afterwards, we draw a plane perpendicular to the intensity axis containing the particular point P. The saturation increases as the point moves away from the intensity axis. Therefore, the distance or projection of the point P from the intensity axis is the saturation (S) associated with the RGB components as shown in Figure 2.21.

The plane we draw passes through the points white, cyan, and black, and any point in the plane has the same hue as shown in Figure 2.21. For any point in the plane, the color components can be specified by linear combination of the points white, cyan, and black. Remember that white contains all the primary colors in equal proportion and black is the absence of color. As the hue is a property of pure color, black and white do not contribute to the hue component of any point in the plane. Hence, the only point that contributes to the hue component is the cyan. Then, all the points lying in the plane have the same hue.

If we rotate 360° in the plane around the intensity axis, we can trace all the possible points that can be specified in the RGB color space to generate all possible hues corresponding to every possible RGB point in the RGB space. In other words, we take the projection of the RGB space in a plane perpendicular to the intensity axis. The vertex of the cube will be projected on a hexagon with the center being the black and white and the colors red, green, and blue projected to different vertices separated by an angle of

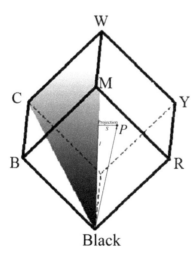

Figure 2.21   Intensity and saturation of a point P in the RGB space.

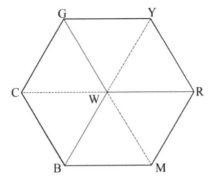

Figure 2.22   Projection perpendicular to the intensity axis.

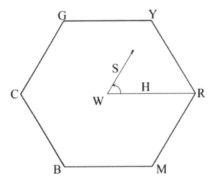

Figure 2.23   Projection of shaded plane in Figure 2.20.

120° among each other as well as yellow, cyan, and magenta as shown in Figure 2.22.

The projection of the shaded plane of Figure 2.21 looks like the straight line S shown in Figure 2.23.

For any point specified in the RGB color space (see the point shown in Figure 2.23), there will be a corresponding point in the projected plane. As we rotate the plane, the particular straight line will be rotated around the center of the hexagon. Therefore, the hue associated with a particular color point in the RGB space is the angle, measured in anticlockwise direction, with respect to the red axis. In other words, the red has a hue of 0° and as we rotate the shaded plane around the intensity axis, the hue will increase. The following formulae are used to convert from RGB to HSI:

$$H = \begin{cases} \theta & \text{if } B \leq G \\ 360^0 - \theta & \text{if } B > G \end{cases}, \tag{2.62}$$

where

$$\theta = \cos^{-1}\left(\frac{0.5\left[(R-G)+(R-B)\right]}{\sqrt{(R-G)^2+(R-B)(G-B)}}\right) \tag{2.63}$$

$$S = 1 - \frac{3}{(R+G+B)}\left[\min\{R,G,B\}\right] \tag{2.64}$$

$$I = \frac{1}{2}\left[R+G+B\right]. \tag{2.65}$$

The inverse conversion formulae are obtained by using different regions

- RG region $(0 \le H < 120°)$

$B = I\,(1-S)$

$$R = I\left[1+\frac{S\cos H}{\cos(60° - H)}\right]$$

$G = 1-(R+B)$

- GB region $(120° \le H < 240°)$, modify H using $H = H - 120°$

$R = I\,(1-S)$

$$G = I\left[1+\frac{S\cos H}{\cos(60° - H)}\right]$$

$B = 1-(R+B)$

- BR region $(240° \le H < 360°)$ modify H using $H = H - 240°$

$G = I\,(1-S)$

$$B = I\left[1+\frac{S\cos H}{\cos(60° - H)}\right]$$

$R = 1-(R+B).$

We have discussed the representation of color that can be present in any image and how to represent those colors in the RGB, CMY, or CMYK and HSI space. Images represented in any of these models can be processed. However, for image compression purposes, the preferred color spaces are

YUV, YCbCr, and YIQ because the luminance component is decoupled from the color components.

The YUV color space is used by the phase alternation line (PAL), National Television System Committee (NTSC), and the Sequentiel Couleur Avec Mémoire or Sequential Color (SECAM) composite color video standards. YUV encodes a color image/video taking into account properties of the human eye that allow for reduced bandwidth for chroma components without perceptual distortion. The equations to convert between RGB and YUV are

$$Y = 0.299R + 0.587G + 0.114B$$
$$U = 0.492\ (B - Y)$$
$$V = 0.877(R - Y).$$

For digital RGB values with a range of 0–255, Y has a range of 0–255, U a range of 0 to ±112, and V a range of 0 to ±157. The inverse equations are

$$R = Y + 1.40V$$
$$G = Y - 0.395U - 0.581V$$
$$B = Y + 2.032U.$$

YIQ color space is derived from the YUV color space and is optionally used by the NTSC composite color video standard. The "I" stands for "inphase" and the "Q" for "quadrature," which is the modulation method used to transmit the color information. Assuming $R,G,B,Y \in [0,1]$, $I \in [-0.5957, 0.5957]$, $Q \in [-0.5226, 0.5226]$, the equations to convert between RGB and YIQ are

$$
\begin{aligned}
Y &= 0.299R + 0.587G + 0.114B \\
I &= 0.596R - 0.275G - 0.321B \\
&= V\cos 33^\circ - U\sin 33^\circ \\
&= 0.736(R - Y) - 0.268(B - Y) \\
Q &= 0.212R' - 0.523G' + 0.311B' \\
&= V\sin 33^\circ + U\cos 33^\circ \\
&= 0.478(R - Y) + 0.413(B - Y)
\end{aligned}
$$

For digital RGB values with a range of 0–255, the component Y has a range of 0–255, the component I has a range from 0 to ±152, and the component Q has a range of 0 to ±134. I and Q are obtained by rotating the U and V axes by 33°. The inverse equations are

$$R = Y + 0.956I + 0.621Q$$
$$G = Y - 0.272I - 0.647Q$$
$$B = Y - 1.107I + 1.704Q.$$

The YCbCr color space is a scaled and offset version of the YUV color space. Y is defined to have a nominal 8-bit range of 16–235; Cb and Cr are defined to have a nominal range of 16–240. There are several YCbCr sampling formats, such as 4:4:4, 4:2:2, 4:1:1, and 4:2:0 that are also described. If the RGB data has a range of 0–255, as is commonly found in computer systems, the following equations to convert between RGB and YCbCr color spaces may be used:

$$Y_{601} = 0.257R + 0.504G + 0.098B + 16$$
$$Cb = -0.148R - 0.291G + 0.439B + 128$$
$$Cr = 0.439R - 0.368G - 0.071B + 128.$$

The inverse equations are

$$R = 1.164(Y_{601} - 16) + 1.596(Cr - 128)$$
$$G = 1.164(Y_{601} - 16) - 0.813(Cr - 128) - 0.391(Cb - 128)$$
$$B = 1.164(Y_{601} - 16) + 2.018(Cb - 128).$$

Other YCbCr color spaces for applications such as high definition television (HDTV) are available, for example, to convert between 8-bit digital RGB data with a 16–235 nominal range and YCbCr are

$$Y_{709} = 0.213R + 0.715G + 0.072B$$
$$Cb = -0.117R - 0.394G + 0.511B + 128$$
$$Cr = 0.511R - 0.464G - 0.047B + 128$$

with inverse equations

$$R = Y_{709} + 1.540(Cr - 128)$$
$$G = Y_{709} - 0.459(Cr - 128) - 0.183(Cb - 128)$$
$$B = Y_{709} + 1.816(Cb - 128).$$

In all the color spaces, for 8-bit sample, the data should be saturated at 0 and 255 to avoid underflow and overflow wrap-around problems. Also, the positioning of the samples is important for image coding.

The YCoCg color model is obtained from a simple and fast transformation of the RGB model. The components obtained are the luma (*Y*), the chrominance orange (Co), and the chrominance green (Cg). Other features are its better decorrelation of the color planes to improve compression performance and exactly lossless invertibility (Rao, 2006). The forward color space transformation is defined by

$$Y = (1/4)R + (1/2)G + (1/4)B$$
$$C_o = (1/2)R + \quad 0 \quad -(1/2)B$$
$$C_g = -(1/4)R + (1/2)G - (1/4)B.$$

The values of *Y* are in the range from 0 to 1, while Co and Cg are in the range of −0.5 to 0.5. The inverse color space transform is

$$R = Y + C_o - C_g$$
$$G = Y + 0 + C_g$$
$$B = Y - C_o - C_g.$$

In the inverse transformation, only two additions and two subtractions are necessary, without real-valued coefficients (Malvar & Sullivan, 2003).

The following are the reversible color transform (RCT) and irreversible color transform (ICT) formulae used in JPEG 2000 (Taubman & Marcellin, 2001). Subscript *r* indicates reversible

Forward RCT:

$$Y_r = \left\lfloor \frac{R + 2G + B}{4} \right\rfloor$$
$$U_r = B - G$$
$$V_r = R - G$$

Inverse RCT:

$$G = Y_r - \left\lfloor \frac{U_r + V_r}{4} \right\rfloor$$
$$R = V_r + G$$
$$B = U_r + G.$$

The ICT is the same as the luminance-chrominance color transformation used in baseline JPEG (Pennebaker & Mitchell, 1993).

Forward ICT:

$$\begin{bmatrix} Y \\ C_b \\ C_r \end{bmatrix} = \begin{bmatrix} 0.299000 & 0.587000 & 0.114000 \\ -0.168736 & -0.331264 & 0.500000 \\ 0.500000 & -0.418688 & -0.081312 \end{bmatrix} \cdot \begin{bmatrix} R \\ G \\ B \end{bmatrix}$$

Inverse ICT:

$$\begin{bmatrix} R \\ G \\ B \end{bmatrix} = \begin{bmatrix} 1.0 & 0.0 & 1.402000 \\ 1.0 & -0.344136 & -0.714136 \\ 1.0 & 1.772000 & 0.0 \end{bmatrix} \cdot \begin{bmatrix} Y \\ C_b \\ C_r \end{bmatrix}.$$

2.8.1 YCbCr Formats

In the 4:4:4 format, each sample has a luminance Y, a blue color component Cb, and a red color component Cr value. The samples are typically 8 bits for consumer applications or 10 bits for pro-video applications per component. Each sample, therefore, requires 24 bits for consumer or 30 bits for pro-video applications. In the 4:2:2 format for every two horizontal Y samples, there is one Cb and one Cr sample. Each sample is typically 8 bits for consumer applications or 10 bits pro-video applications per component. Therefore, this format requires 16 bits for consumer or 20 bits for pro-video applications.

The 4:4:1 format is also known as the YUV12 format. It is specially used in consumer video and DV video compression applications. For every four horizontal Y samples, there is one Cb and Cr value. Each component is typically 8 bits. Each sample, therefore, requires 12 bits. In the 4:2:0 format, YCbCr has a 2:1 reduction of Cb and Cr in both the vertical and horizontal directions rather than the horizontal-only 2:1 reduction of Cb and Cr used in 4:2:2. The 4:2:0 is commonly used for video compression. To display 4:2:2, 4:1:1, and 4:2:0 YCbCr data, they are first converted to 4:4:4 YCbCr data, using interpolation to generate the new Cb and Cr samples. Figure 2.24 illustrates the positioning of YCbCr samples for the (a) 4:4:4, (b) 4:2:2, (c) 4:1:1 co-sited sampling, and (d) and (e) 4:2:0 sampling for H.261, H.263, MPEG1, and MPEG 2 progressive and interlaced pictures.

2.9 Chromaticity Diagram

The color gamut perceived by a person with normal vision is shown in Figure 2.25. Color perception was carried out using the primary colors: red with a 700 nm wavelength, green at 546.1 nm, and blue at 435.8 nm. The primary

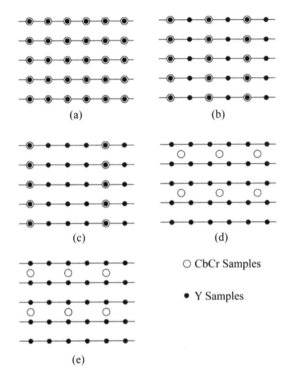

Figure 2.24   Positioning of YCbCr samples for the (a) 4:4:4, (b) 4:2:2, (c) 4:1:1 co-sited sampling, (d) 4:2:0 for H.261, H.263 MPEG 1, and (e) 4:2:0 for MPEG 2 sampling formats.

and other spectrally pure colors, resulting from mixing the primary colors, are placed along the boundary line of the curve. Note that the color purple is a combination of red and blue colors. Therefore, the purple line connects these two colors at the end of the spectrum. When a color approaches the boundary, it becomes more saturated.

In the International Commission on Illumination (CIE) system, the intensities of red, green, and blue are transformed into what are called the tristimulus values (X, Y, Z). The lower case letters x, y, and z are called chromaticity coordinates of Figure 2.25 and add up to 1 (Masaoka, 2016), (Jack, 2008). These are the coordinates calculated as

$$x = red/(red + green + blue) \quad = X/(X + Y + Z)$$
$$y = green/(red + green + blue) = Y/(X + Y + Z)$$
$$z = blue/(red + green + blue) = Z/(X + Y + Z).$$

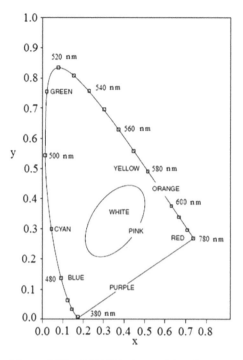

Figure 2.25 CIE 1931 chromaticity diagram.

The chromaticities of the RGB primary colors and reference white (CIE illuminate D_{65}) for modern NTSC systems (Masaoka, 2016), (Jack, 2008) are

$$R: \quad x_r = 0.630 \quad y_r = 0.340$$
$$G: \quad x_g = 0.310 \quad y_g = 0.595$$
$$B: \quad x_b = 0.155 \quad y_b = 0.070$$
$$W: \quad x_w = 0.3127 \quad y_w = 0.3290$$

The chromaticities of the RGB primary colors and reference white (CIE illuminate D_{65}) for PAL and NTSC systems (Masaoka, 2016), (Jack, 2008) are

$$R: \quad x_r = 0.64 \quad y_r = 0.33$$
$$G: \quad x_g = 0.29 \quad y_g = 0.6$$
$$B: \quad x_b = 0.15 \quad y_b = 0.06$$
$$W: \quad x_w = 0.3127 \quad y_w = 0.3290$$

The chromaticities of the RGB primary colors and reference white (CIE illuminate D_{65}) for HDTV systems are (Jack, 2008)

$$R: \quad x_r = 0.64 \qquad y_r = 0.33$$
$$G: \quad x_g = 0.3 \qquad y_g = 0.6$$
$$B: \quad x_b = 0.15 \qquad y_b = 0.06$$
$$W: \quad x_w = 0.3127 \quad y_w = 0.3290$$

The coordinate z can be expressed in terms of x and y. Hence, only x and y are required to specify any color. The color gamut that a source or display can handle is specified by three sets of coordinates (x,y) to form a triangle inside the chromaticity diagram. NTSC, PAL, SECAM, inks/dyes, etc. specify these coordinates for their color gamut.

3

The Discrete Cosine and Sine Transforms

Transform coding has been widely studied and used in image and video compression because it has the following characteristics:

- The optimal transform completely decorrelates the data in a sequence/block; i.e., it packs the most amount of energy in the fewest number of coefficients. In this way, many coefficients can be discarded after quantization and prior to encoding. It is important to note that the transform operation itself does not achieve any compression. It aims at decorrelating the original data and compacting a large fraction of the signal energy into relatively few transform coefficients.
- Owing to the large statistical variations among data, the optimum transform usually depends on the data, and finding the basis functions of such transform is a computationally intensive task. This is particularly a problem if the data blocks are highly non-stationary, which necessitates the use of more than one set of basis functions to achieve high decorrelation. Therefore, it is desirable to trade optimum performance for a transform whose basis functions are data-independent.
- The number of operations required for an N-point transform is generally of the order $O(n^2)$. Some transforms have fast implementations, which reduce the number of operations to $O(n\log_2 n)$. For a separable $n \times n$ 2D transform, performing the row and column 1D transforms successively reduces the number of operations from $O(n^4)$ to $O(2n^2 \log_2 n)$.

The discrete cosine (DCT) and discrete sine (DST) transforms are members of a family of sinusoidal unitary transforms (Britanak, Yip, & Rao, 2006). A sinusoidal unitary transform is an invertible linear transform whose kernel is defined by a set of complete, orthogonal/orthonormal discrete cosine and/or sine basis functions. They are having a great relevance to data compression. Subbarayappa, S. and Aradhyamath, P.G., 2021, April. Analytical Transform for Image Compression. In 2021 6th IEEE International Conference for Convergence in Technology (I2CT) (pp. 1–5).

3.1 The DCT

Inversion is part of transform processing. In practice, if the non-singular matrix is real and orthogonal, its inverse is easily obtained as its transpose. Such unitary transform matrices (with the proper normalizations) have a preeminent place in the digital signal-processing field. We present the definitions for the four discrete cosine transform kernels as classified by Wang (Wang, 1984).

I. DCT-I. For the order $N + 1$.

$$\left[C_{N+1}^{I} \right]_{m\,n} = \sqrt{\frac{2}{N}} \left[k_m k_n \cos\left(\frac{\pi m\, n}{N} \right) \right] \qquad m, n = 0, 1, \ldots, N \qquad (3.1)$$

and: $\left[C_N^{I} \right]^{T} = \left[C_N^{I} \right]^{-1} = \left[C_N^{VI} \right] \Rightarrow \left[C_N^{V} \right]\left[C_N^{VI} \right] = \mathbf{I}.$

II. DCT-II. Excellent energy compaction property and best approximation to Karhunen–Loève transform (KLT).

$$\left[C_N^{II} \right]_{m\,n} = \sqrt{\frac{2}{N}} \left[k_n \cos\left(\frac{\pi\, (m + \frac{1}{2})n}{N} \right) \right] \qquad m, n = 0, 1, \ldots, N - 1. \quad (3.2)$$

and: $\left[C_N^{II} \right]^{T} = \left[C_N^{II} \right]^{-1} = \left[C_N^{III} \right] \Rightarrow \left[C_N^{II} \right]\left[C_N^{III} \right] = \mathbf{I}.$

III. DCT-III. Inverse DCT-II

$$\left[C_N^{III} \right]_{mn} = \sqrt{\frac{2}{N}} \left[k_m \cos\left(\frac{\pi\, m(n + \frac{1}{2})}{N} \right) \right] \qquad m, n = 0, 1, \ldots, N - 1. \quad (3.3)$$

and: $\left[C_N^{III} \right]^{T} = \left[C_N^{III} \right]^{-1} = \left[C_N^{II} \right] \Rightarrow \left[C_N^{III} \right]\left[C_N^{II} \right] = \mathbf{I}.$

IV. DCT-IV. Fast implementation of lapped orthogonal transform for the efficient transform-subband coding.

$$\left[C_N^{IV} \right]_{mn} = \sqrt{\frac{2}{N}} \left[\cos\left(\frac{\pi\, (m + \frac{1}{2})(n + \frac{1}{2})}{N} \right) \right] \qquad m, n = 0, 1, \ldots, N - 1. \quad (3.4)$$

and: $\left[C_N^{IV} \right]^{T} = \left[C_N^{IV} \right]^{-1} = \left[C_N^{IV} \right] \Rightarrow \left[C_N^{IV} \right]\left[C_N^{IV} \right] = \mathbf{I}$

$$\text{where: } k_p = \begin{cases} 1 & \text{if } p \neq 0 \text{ or } p \neq N \\ \frac{1}{\sqrt{2}} & \text{if } p = 0 \text{ or } p = N \end{cases}.$$

Here, we note that DCT-II is the discrete cosine transform first developed by Ahmed, Natarajan, and Rao (1974). The DCT-III is obviously the transpose of the DCT-II, and the DCT-IV is the shifted version of DCT-I. There are four more combinations of the discrete cosine transform. DCT types I–IV are equivalent to real-even discrete Fourier transforms (DFTs) of even order; therefore, there are four additional types of DCT corresponding to real-even DFTs of odd order as shown in the following equations (Britanak & Rao, 2007):

$$\text{V. } \left[C_N^V \right]_{m\,n} = \frac{2}{\sqrt{2N-1}} \left[k_m k_n \cos\left(\left(\frac{2\pi}{2N-1} \right) m\,n \right) \right] \qquad m, n = 0,1,\ldots,N-1$$

$$(3.5)$$

$$\text{and: } \left[C_N^V \right] = \left[C_N^V \right]^T = \left[C_N^V \right]^{-1} \Rightarrow \left[C_N^V \right]\left[C_N^V \right] = \mathbf{I}$$

$$\text{where } k_m, k_n = \begin{cases} 1 & \text{if } m,n \neq 0 \\ \frac{1}{\sqrt{2}} & \text{if } m,n = 0 \end{cases}.$$

$$\text{VI. } \left[C_N^{VI} \right]_{m\,n} = \frac{2}{\sqrt{2N-1}} \left[k_m k_n \cos\left(\left(\frac{2\pi}{2N-1} \right) (m+\tfrac{1}{2})n \right) \right], \qquad m, n = 0,1,\ldots,N-1$$

$$(3.6)$$

$$\text{and: } \left[C_N^{VI} \right]^T = \left[C_N^{VI} \right]^{-1} = \left[C_N^{VII} \right] \Rightarrow \left[C_N^{VI} \right]\left[C_N^{VII} \right] = \mathbf{I}$$

$$\text{where: } k_m = \begin{cases} 1 & \text{if } m \neq N-1 \\ \frac{1}{\sqrt{2}} & \text{if } m = N-1 \end{cases}$$

$$k_n = \begin{cases} 1 & \text{if } n \neq 0 \\ \frac{1}{\sqrt{2}} & \text{if } n = 0 \end{cases}.$$

$$\text{VII. } \left[C_N^{VII} \right]_{m\,n} = \frac{2}{\sqrt{2N-1}} \left[k_n k_m \cos\left(\left(\frac{2\pi}{2N-1} \right) m\left(n+\tfrac{1}{2}\right) \right) \right], \qquad m, n = 0,1,\ldots,N-1$$

$$(3.7)$$

$$\text{and: } \left[C_N^{VII} \right]^T = \left[C_N^{VII} \right]^{-1} = \left[C_N^{VI} \right] \Rightarrow \left[C_N^{VII} \right]\left[C_N^{VI} \right] = \mathbf{I}$$

$$\text{where: } k_m = \begin{cases} 1 & \text{if } m \neq 0 \\ \frac{1}{\sqrt{2}} & \text{if } m = 0 \end{cases}$$

$$k_n = \begin{cases} 1 & \text{if } n \neq N-1 \\ \frac{1}{\sqrt{2}} & \text{if } n = N-1 \end{cases}.$$

$$\text{VIII. } \left[C_N^{VIII} \right]_{m\,n} = \frac{2}{\sqrt{2N+1}} \left[\cos\left(\left(\frac{2\pi}{2N+1} \right)(m+\tfrac{1}{2})(n+\tfrac{1}{2}) \right) \right], \quad m,n = 0,1,\ldots,N-1$$

$$\text{and: } \left[C_N^{VIII} \right] = \left[C_N^{VIII} \right]^T = \left[C_N^{VIII} \right]^{-1} \Rightarrow \left[C_N^{VIII} \right]\left[C_N^{VIII} \right] = \mathbf{I} \tag{3.8}$$

where $\mathbf{I}$ is the identity matrix and $[\bullet]^T$ stands for transposed matrix. It should be noted that aside from the scaling factors, DCT-III is the transpose of DCT-II and DCT-VI is the transpose of DCT-VII. Those transform matrices that have a flat vector are specially adapted to analyzing signals with DC components. On the other hand, DCT-IV and DCT-VIII "reach" beyond the boundary and are good candidates as "lapped transforms."

The objective of the transformation coding is to represent a signal in a set of waveforms, each with a particular spatial frequency, in such a way that the structures that are perceptible by the human eye are separated from the rest of the structures to assign more importance to the perceivable. The DCT provides this separation by its basis functions. Figure 3.1 shows the basis function waveforms of the DCT-II. The waveforms are orthogonal and linearly independent so that they can represent any eight-sample values. This representation is known as the DCT coefficients.

The coefficient that scales the constant basis function C_0 is called the DC coefficient and the rest are called the AC coefficients.

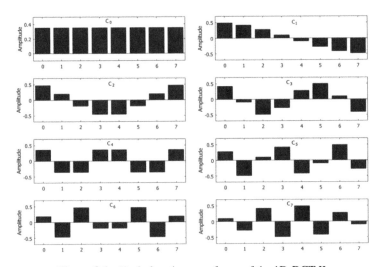

Figure 3.1   Basis function waveforms of the 1D-DCT-II.

3.2 The DST

Several relationships among the DCT-II, DCT-III, and the DCT-IV have been studied, leading to low-complexity factorizations for this kind of transforms (Bi & Zeng, 2003), (Chen, Smith, & Fralick, 1977), (Wang, 1984), (Vetterli & Nussbaumer, 1984), (Lee, 1984), (Loeffler, Ligtenberg, & Moschytz, 1989), (Hou, 1987). Recently, Saxena *et al.* (Han, Saxena, & Rose, 2010), (Saxena & Fernandes, 2013) showed that transforms other than the DCT-II can better represent signals produced by the intra-prediction of an image/video coding scheme. In particular, they demonstrated that the discrete sine transform of type-VII (DST-VII) approaches the optimal KLT in terms of decorrelation for intra-predicted signals. For this reason, the DST-VII was specified in the high efficiency video coding (HEVC) standard to code intra-predicted 4×4 luminance blocks. The DST-I matrices are (Britanak & Rao, 2007) as follows:

I. DST-I.

$$\left[S_{N-1}^{I} \right]_{m\,n} = \sqrt{\frac{2}{N}} \left[\sin\left(\frac{\pi m n}{N} \right) \right] \qquad m, n = 1,\ldots,N-1. \qquad (3.9)$$

II. DST-II.

$$\left[S_{N}^{II} \right]_{m\,n} = \sqrt{\frac{2}{N}}\, k_n \left[\sin\left(\frac{\pi \left(m+\frac{1}{2}\right)(n+1)}{N} \right) \right], \qquad m, n = 0,1,\ldots,N-1. \qquad (3.10)$$

III. DST-III.

$$\left[S_{N}^{III} \right]_{m\,n} = \sqrt{\frac{2}{N}}\, k_m \left[\sin\left(\frac{\pi \left(m+1\right)\left(n+\frac{1}{2}\right)}{N} \right) \right], \qquad m, n = 0,1,\ldots,N-1. \qquad (3.11)$$

IV. DST-IV.

$$\left[S_{N}^{IV} \right]_{m\,n} = \sqrt{\frac{2}{N}} \left[\sin\left(\frac{\pi \left(m+\frac{1}{2}\right)\left(n+\frac{1}{2}\right)}{N} \right) \right], \qquad m, n = 0,1,\ldots,N-1. \qquad (3.12)$$

V. DST-V.

$$\left[S_{N-1}^{V} \right]_{m\,n} = \frac{2}{\sqrt{2N-1}} \left[\sin\left(\frac{2\pi m n}{2N-1} \right) \right] \qquad m, n = 1,\ldots,N-1. \qquad (3.13)$$

VI. DST-VI.

$$\left[S_{N-1}^{VI} \right]_{m\,n} = \frac{2}{\sqrt{2N-1}} \left[\sin\left(\frac{2\pi \left(m+\frac{1}{2}\right)(n+1)}{2N-1} \right) \right] \qquad m, n = 0,1,\ldots,N-2. \qquad (3.14)$$

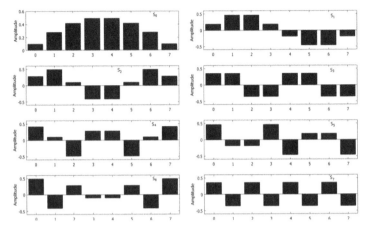

Figure 3.2   Basis function waveforms of the 1D-DST-II.

VII. DST-VII.

$$\left[S_{N-1}^{VII}\right]_{m\,n} = \frac{2}{\sqrt{2N-1}}\left[\sin\left(\frac{2\pi(m+1)\left(n+\frac{1}{2}\right)}{2N-1}\right)\right] \qquad m,n = 0,1,\ldots,N-2.$$

(3.15)

VIII. DST-VIII.

$$\left[S_{N}^{VII}\right]_{m\,n} = \frac{2}{\sqrt{2N-1}}k_{m}k_{n}\left[\sin\left(\frac{2\pi\left(m+\frac{1}{2}\right)\left(n+\frac{1}{2}\right)}{2N-1}\right)\right] \qquad m,n = 0,1,\ldots,N-1$$

(3.16)

$$\text{where: } k_{l} = \begin{cases} 1 & \text{if } l \neq 0 \\ \frac{1}{\sqrt{2}} & \text{if } l = N-1 \end{cases}.$$

Figure 3.2 shows the basis function waveforms of the 1D-DST-II.

3.3 The Fast DCT

A fast transform is a sequence of operations in the form of weighted sums and differences, which efficiently computes the coefficients of the transform with a reduced number of operations. The fast DCT structure takes advantage of the symmetries of the basis functions (see Figure 3.1 and Equation (3.17)) to reduce the number of operations. The methods convert the DCT matrix into an alternative transformation matrix that can be decomposed into sparse matrices of low complexity.

$$
[C_8^{II}] = \begin{bmatrix}
0.3536 & 0.3536 & 0.3536 & 0.3536 & 0.3536 & 0.3536 & 0.3536 & 0.3536 \\
0.4904 & 0.4157 & 0.2778 & 0.0975 & -0.0975 & -0.2778 & -0.4157 & -0.4904 \\
0.4619 & 0.1913 & -0.1913 & -0.4619 & -0.4619 & -0.1913 & 0.1913 & 0.4619 \\
0.4157 & -0.0975 & -0.4904 & -0.2778 & 0.2778 & 0.4904 & 0.0975 & -0.4157 \\
0.3536 & -0.3536 & -0.3536 & 0.3536 & 0.3536 & -0.3536 & -0.3536 & 0.3536 \\
0.2778 & -0.4904 & 0.0975 & 0.4157 & -0.4157 & -0.0975 & 0.4904 & -0.2778 \\
0.1913 & -0.4619 & 0.4619 & -0.1913 & -0.1913 & 0.4619 & -0.4619 & 0.1913 \\
0.0975 & -0.2778 & 0.4157 & -0.4904 & 0.4904 & -0.4157 & 0.2778 & -0.0975
\end{bmatrix}.
$$

$$(3.17)$$

The transformed sequence can be represented in vector notations as follows:

$$
\mathbf{Y}_8^{II} = \left(\sqrt{\frac{2}{N}} \right) [C_8^{II}] \mathbf{x}. \tag{3.18}
$$

Here, $[C_8^{II}]$ is a $(N \times N)$ DCT-II matrix. For the sake of visual clarity, the superscript II will be omitted. When $[C_8]$ is decomposed into sparse matrices, a fast algorithm results as follows (Rao & Yip, 1990):

$$
[C_8] = \mathbf{A}_2 \mathbf{A}_4 \tag{3.19}
$$

where

$$
\mathbf{A}_2 = \begin{bmatrix} \dfrac{1}{\sqrt{2}} & \dfrac{1}{\sqrt{2}} \\ c_4^1 & c_4^3 \end{bmatrix}. \tag{3.20}
$$

Here, $c_k^i = \cos\left(\pi \dfrac{i}{k} \right)$. And

$$
\mathbf{A}_4 = \begin{bmatrix} 1 & 0 & 0 & 0 \\ 0 & 0 & 0 & 1 \\ 0 & 1 & 0 & 0 \\ 0 & 0 & 1 & 0 \end{bmatrix} \begin{bmatrix} \dfrac{1}{\sqrt{2}} & \dfrac{1}{\sqrt{2}} & \dfrac{1}{\sqrt{2}} & \dfrac{1}{\sqrt{2}} \\ c_4^1 & c_4^3 & c_4^3 & c_4^1 \\ c_8^3 & -c_8^1 & c_8^1 & -c_8^3 \\ c_8^1 & c_8^3 & -c_8^3 & -c_8^1 \end{bmatrix} =
$$

$$
\begin{bmatrix} 1 & 0 & 0 & 0 \\ 0 & 0 & 0 & 1 \\ 0 & 1 & 0 & 0 \\ 0 & 0 & 1 & 0 \end{bmatrix}
\begin{bmatrix} \dfrac{1}{\sqrt{2}} & \dfrac{1}{\sqrt{2}} & 0 & 0 \\ c_4^1 & c_4^3 & 0 & 0 \\ 0 & 0 & -c_8^1 & c_8^3 \\ 0 & 0 & c_8^3 & c_8^1 \end{bmatrix}
\begin{bmatrix} 1 & 0 & 0 & 1 \\ 0 & 1 & 1 & 0 \\ 0 & 1 & -1 & 0 \\ 1 & 0 & 0 & -1 \end{bmatrix}.
$$

$$(3.21)$$

Observe that $\mathbf{A}_4$ can be expressed as

$$
\mathbf{A}_4 = \mathbf{P}_4 \begin{bmatrix} \mathbf{A}_2 & \mathbf{0} \\ \mathbf{0} & \mathbf{R}_2 \end{bmatrix} \mathbf{B}_4,
\qquad (3.22)
$$

where $\mathbf{P}_4$ is the (4×4) permutation matrix, $\mathbf{B}_4$ is the (4×4) butterfly matrix, and $\overline{\mathbf{R}}_2$ is the (2×2) remaining matrix. In general, for N being a power of 2, Equation (3.22) is in the form

$$
\mathbf{A}_N = \mathbf{P}_N \begin{bmatrix} \mathbf{A}_{N/2} & \mathbf{0} \\ \mathbf{0} & \overline{\mathbf{R}}_{N/2} \end{bmatrix} \mathbf{B}_N.
\qquad (3.23)
$$

$\mathbf{P}_N$ permutes the even rows in increasing order in the upper half and the odd rows in decreasing order in the lower half. $\mathbf{B}_N$ can be expressed in terms of the identity matrix $\mathbf{I}_{N/2}$ and the opposite identity matrix $\mathbf{J}_{N/2}$ as follows:

$$
\mathbf{B}_N = \begin{bmatrix} \mathbf{I}_{N/2} & \mathbf{J}_{N/2} \\ \mathbf{J}_{N/2} & -\mathbf{I}_{N/2} \end{bmatrix}.
\qquad (3.24)
$$

The matrix $\overline{\mathbf{R}}$ is obtained by reversing the orders of both the rows and columns of the matrix $\mathbf{R}_N$ whose ik-element is given by

$$
[\mathbf{R}_N]_{ik} = \cos\left(\frac{(2i+1)(2k+1)}{4N} \pi \right), \quad i,k = 0,1,...,N-1.
\qquad (3.25)
$$

$[\mathbf{R}_N]_{ik}$ defines the transform matrix for a DCT of type IV (DCT-IV). $\mathbf{A}_N$ is partly recursive since the matrix $\mathbf{R}_N$ is not recursive.

Feig and Winogard (1992) proposed an efficient 8-point DCT algorithm that requires 11 multiplications and 29 additions base in the factorization of matrix 3.17. This matrix can be expressed as

$$
[C_8^{II}] = \mathbf{P}_8 \mathbf{K}_8 \mathbf{B}_1 \mathbf{B}_2 \mathbf{B}_3
\qquad (3.26)
$$

where

$$\mathbf{P}_8 = \begin{bmatrix} 1 & 0 & 0 & 0 & 0 & 0 & 0 & 0 \\ 0 & 0 & 0 & 0 & -1 & 0 & 0 & 0 \\ 0 & 0 & 1 & 0 & 0 & 0 & 0 & 0 \\ 0 & 0 & 0 & 0 & 0 & -1 & 0 & 0 \\ 0 & 1 & 0 & 0 & 0 & 0 & 0 & 0 \\ 0 & 0 & 0 & 0 & 0 & 0 & 0 & -1 \\ 0 & 0 & 0 & 1 & 0 & 0 & 0 & 0 \\ 0 & 0 & 0 & 0 & 0 & 0 & 1 & 0 \end{bmatrix}$$

$\mathbf{K}_8$ is a block diagonal matrix

$$\mathbf{K}_8 = \frac{1}{2}\text{diag}(\mathbf{G}_1,\mathbf{G}_1,\mathbf{G}_2,\mathbf{G}_4)$$

$$\mathbf{G}_1 = \cos(\pi/4)$$

$$\mathbf{G}_2 = \begin{bmatrix} \cos(3\pi/8) & \cos(\pi/8) \\ -\cos(\pi/8) & \cos(3\pi/8) \end{bmatrix}$$

$$\mathbf{G}_4 = \begin{bmatrix} \cos(5\pi/16) & \cos(9\pi/16) & \cos(3\pi/16) & \cos(\pi/16) \\ -\cos(\pi/16) & \cos(5\pi/16) & \cos(9\pi/16) & \cos(3\pi/16) \\ -\cos(3\pi/16) & -\cos(\pi/16) & \cos(5\pi/16) & \cos(9\pi/16) \\ -\cos(9\pi/16) & -\cos(3\pi/16) & -\cos(\pi/16) & \cos(5\pi/16) \end{bmatrix}$$

and

$$B_1 = \begin{bmatrix} 1 & 1 & 0 & 0 & 0 & 0 & 0 & 0 \\ 1 & -1 & 0 & 0 & 0 & 0 & 0 & 0 \\ 0 & 0 & 0 & 1 & 0 & 0 & 0 & 0 \\ 0 & 0 & 1 & 0 & 0 & 0 & 0 & 0 \\ 0 & 0 & 0 & 0 & 0 & 0 & -1 & 0 \\ 0 & 0 & 0 & 0 & 0 & 0 & 0 & 1 \\ 0 & 0 & 0 & 0 & 0 & -1 & 0 & 0 \\ 0 & 0 & 0 & 0 & -1 & 0 & 0 & 0 \end{bmatrix}, \quad B_2 = \begin{bmatrix} 1 & 0 & 0 & 1 & 0 & 0 & 0 & 0 \\ 0 & 1 & 1 & 0 & 0 & 0 & 0 & 0 \\ 1 & 0 & 0 & -1 & 0 & 0 & 0 & 0 \\ 0 & 1 & -1 & 0 & 0 & 0 & 0 & 0 \\ 0 & 0 & 0 & 0 & 0 & 1 & 0 & 0 \\ 0 & 0 & 0 & 0 & 0 & 0 & 1 & 0 \\ 0 & 0 & 0 & 0 & 0 & 0 & 0 & 1 \end{bmatrix}, \quad B_3 = \begin{bmatrix} 1 & 0 & 0 & 0 & 0 & 0 & 0 & 1 \\ 0 & 1 & 0 & 0 & 0 & 0 & 1 & 0 \\ 0 & 0 & 1 & 0 & 0 & 1 & 0 & 0 \\ 0 & 0 & 0 & 1 & 1 & 0 & 0 & 0 \\ 1 & 0 & 0 & 0 & 0 & 0 & 0 & -1 \\ 0 & 1 & 0 & 0 & 0 & 0 & -1 & 0 \\ 0 & 0 & 1 & 0 & 0 & -1 & 0 & 0 \\ 0 & 0 & 0 & 1 & -1 & 0 & 0 & 0 \end{bmatrix}.$$

3.4 Structures for Implementing the Fast 1-D DCT

Vetterli and Ligtenberg (1986) developed an 8-point fast DCT algorithm and chip for a data rate of 100 Mbits/s. The signal flowgraph shown in Figure 3.3 is a modified version (Rao & Yip, 1990) of the proposed signal flowgraph by (Vetterli & Ligtenberg, 1986) and consists of five stages where standard rotations $R(\alpha)$ are performed in output stage.

The rotator is given by

$$\begin{matrix} a_i \to \\ a_k \to \end{matrix} \left[R(\alpha) \right] \begin{matrix} \to b_i \\ \to b_k \end{matrix} \quad \Rightarrow \quad \begin{bmatrix} b_i \\ b_k \end{bmatrix} = \begin{bmatrix} \cos(\alpha) & \sin(\alpha) \\ -\sin(\alpha) & \cos(\alpha) \end{bmatrix} \begin{bmatrix} a_i \\ a_k \end{bmatrix}.$$

Figure 3.4 shows the signal flowgraph for the 8-point fast (scaled) 1D AAN DCT algorithm proposed by Arai, Argui, and Nakajima (1988). The algorithm adopts the small and fast Fourier transform (FFT) developed in (Winograd, 1978) requiring only five multiplications and 29 additions. Outputs of the flowgraph should be multiplied by constants. However, the multiplications can be absorbed into the quantization stage giving and overall computation reduction. To obtain the fast inverse DCT (IDCT), the graph needs to be reversed (i.e., read from right to left and reverse the arrows).

The DCT computation may also be examined through other transforms. For example, via the discrete Hartley transform, as shown in Figure 3.5,

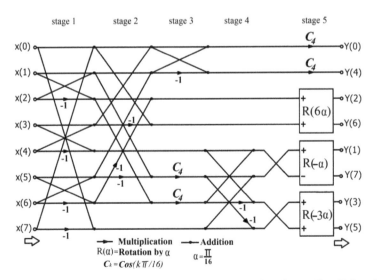

Figure 3.3   Signal flowgraph of the 8-point, forward 1D-DCT of Vetterli and Ligtenberg.

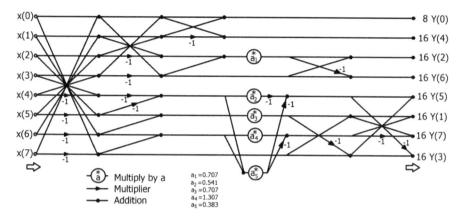

Figure 3.4 Signal flowgraph of the 8-point, fast (scaled) AAN forward 1D-DCT.

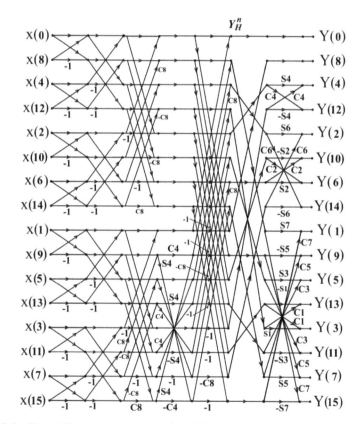

Figure 3.5 Signal flowgraph for computing DCT via DHT. $C_k = \cos(k\pi/32)$ and $S_k = \sin(k\pi/32)$, $N = 16$.

because it has a close relationship to the Fourier transform and its apparent advantage in handling real data. The amount of computation necessary for an N-point sequence compares favorably with those of other algorithms.

In addition to the algorithms described, there are many more. Some are developed for specific applications. Others are developed for specific numbers of input sample points. Many algorithms have been considered for VLSI implementation. In such an implementation, the orderly architecture and the numerical stability are important considerations in addition to the recursive and modular nature of the algorithm.

3.5 2D DCT

For a given 2D data sequence $\{x(i, k): \ i,k = 0,1,\ldots,N-1\}$, the 2D DCT sequence $\{Y(m,n): \ m,n = 0,1,\ldots,N-1\}$ is given by

$$Y(m,n) = \frac{1}{N^2} k_m k_n \sum_{i=0}^{N-1} \sum_{k=0}^{N-1} x(i,k) \left[\cos\frac{(2i+1)m\pi}{2N} \right] \left[\cos\frac{(2k+1)n\pi}{2N} \right].$$

(3.27)

The array in the data domain is

$$x(i,k) = \frac{1}{N^2} \sum_{m=0}^{N-1} \sum_{n=0}^{N-1} k_m k_n Y(m,n) \left[\cos\frac{(2i+1)m\pi}{2N} \right] \left[\cos\frac{(2k+1)n\pi}{2N} \right]$$

(3.28)

$$\text{where } k_p = \begin{cases} \dfrac{1}{\sqrt{2}} & \text{for } p = 0, \\ 1 & \text{otherwise.} \end{cases}$$

Equation (3.27) can be written in matrix form as

$$[\mathbf{Y}_L] = [\mathbf{D}(N,N)][\mathbf{x}_L]$$

(3.29)

where $[\mathbf{Y}_L]$ and $[\mathbf{X}_L]$ are sequences rearranged in lexicographic order of the 2D transformed and data arrays respectively, i.e.,

$$[\mathbf{Y}_L] = \{Y(0,0), Y(0,1), \ldots, Y(0,N-1), Y(1,0), Y(1,1), \ldots, Y(N-1,N-1)\}$$

$$[\mathbf{x}_L] = \{x(0,0), x(0,1), \ldots, x(0,N-1), x(1,0), x(1,1), \ldots, x(N-1,N-1)\}.$$

$[\mathbf{D}(N,N)_L]$ is an $(N^2 \times N^2)$ matrix defined as

$$[\mathbf{D}(N,N)] = [\mathbf{L}(N)] \otimes [\mathbf{L}(N)].$$

(3.30)

$[\mathbf{L}(N)]$ is the $(N \times N)$ 1D DCT matrix and $\otimes$ is the Kronecker product. For example,

$$[\mathbf{L}(4)] = \begin{bmatrix} c_2 & c_2 & c_2 & c_2 \\ c_1 & c_3 & c_5 & c_7 \\ c_2 & c_6 & c_{10} & c_{14} \\ c_3 & c_9 & c_{15} & c_{21} \end{bmatrix},$$

(3.31)

$c_i = \cos\left(\dfrac{i\,\pi}{8}\right)$. Application of the trigonometric identities to Equation (3.31) will result in

$$[\mathbf{D}(4,4)] = \begin{bmatrix} c_{2,2} & c_{2,2} & c_{2,2} & c_{2,2} & c_{2,2} & c_{2,2} & c_{2,2} & c_{2,2} \\ c_{2,1} & c_{2,3} & -c_{2,3} & -c_{2,1} & c_{2,1} & c_{2,3} & -c_{2,3} & -c_{2,1} \\ c_{2,2} & -c_{2,2} & -c_{2,2} & c_{2,2} & c_{2,2} & -c_{2,2} & -c_{2,2} & c_{2,2} \\ c_{2,3} & -c_{2,1} & c_{2,1} & -c_{2,3} & c_{2,3} & -c_{2,1} & c_{2,1} & -c_{2,3} \\ c_{1,2} & c_{1,2} & c_{1,2} & c_{1,2} & c_{3,2} & c_{3,2} & c_{3,2} & c_{3,2} \\ c_{1,1} & c_{1,3} & -c_{1,3} & -c_{1,1} & c_{3,1} & c_{3,3} & -c_{3,3} & -c_{3,1} \\ c_{1,2} & -c_{1,2} & -c_{1,2} & c_{1,2} & c_{3,2} & -c_{3,2} & -c_{3,2} & c_{3,2} \\ c_{1,3} & -c_{1,1} & c_{1,1} & -c_{1,3} & c_{3,3} & -c_{3,1} & c_{3,1} & -c_{3,3} \\ c_{2,2} & c_{2,2} & c_{2,2} & c_{2,2} & -c_{2,2} & -c_{2,2} & -c_{2,2} & -c_{2,2} \\ c_{2,1} & c_{2,3} & -c_{2,3} & -c_{2,1} & -c_{2,1} & -c_{2,3} & c_{2,3} & c_{2,1} \\ c_{2,2} & -c_{2,2} & -c_{2,2} & c_{2,2} & -c_{2,2} & c_{2,2} & c_{2,2} & -c_{2,2} \\ c_{2,3} & -c_{2,1} & c_{2,1} & -c_{2,3} & -c_{2,3} & c_{2,1} & -c_{2,1} & c_{2,3} \\ c_{3,2} & c_{3,2} & c_{3,2} & c_{3,2} & -c_{1,2} & -c_{1,2} & -c_{1,2} & -c_{1,2} \\ c_{3,1} & c_{3,3} & -c_{3,3} & -c_{3,1} & -c_{1,1} & -c_{1,3} & c_{1,3} & c_{1,1} \\ c_{3,2} & -c_{3,2} & -c_{3,2} & c_{3,2} & -c_{1,2} & c_{1,2} & c_{1,2} & -c_{1,2} \\ c_{3,3} & -c_{3,1} & c_{3,1} & -c_{3,3} & -c_{1,3} & c_{1,1} & -c_{1,1} & c_{1,3} \end{bmatrix} \begin{matrix} Even \\ Odd \\ Even \\ Odd \\ Even \\ Odd \\ Even \\ Odd \\ Even \\ Odd \\ Even \\ Odd \\ Even \\ Odd \\ Even \\ Odd \end{matrix}$$

(3.32)

where $c_{u,v} = \cos\left(\dfrac{u\,\pi}{2N}\right)\cos\left(\dfrac{v\,\pi}{2N}\right)$. In Equation (3.32), only the first half of the elements of each row vector are shown. Note the symmetry of the row

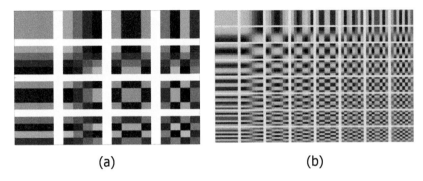

(a)   (b)

Figure 3.6   Basis images for (a) 4×4 size and (b) 8×8 size.

vectors. Figure 3.6 shows the 2D DCT basis images for (a) 4×4 size and (b) 8×8 size.

3.6 Hierarchical Lapped Transform

In the work proposed by Tran, Liang, and Tu (2003), each overlap-filtering operator is centered between the boundaries of four core transform operators. The four-channel time-domain lapped transform (TDLT) is defined as (Suzuki & Yoshida, 2017)

$$\mathbf{E}(\mathbf{z}) = \mathbf{P}_4 \overbrace{\begin{bmatrix} \mathbf{R}_{\frac{\pi}{4}} & \mathbf{0} \\ \mathbf{0} & \mathbf{R}_{\frac{\pi}{8}}\mathbf{J}_2 \end{bmatrix}}^{\text{core transform}} \mathbf{W}_4 \begin{bmatrix} \mathbf{I}_2 & \mathbf{0} \\ \mathbf{0} & z^{-1}\mathbf{I}_2 \end{bmatrix} \begin{bmatrix} \mathbf{0} & \mathbf{I}_2 \\ \mathbf{I}_2 & \mathbf{0} \end{bmatrix} \mathbf{W}_4 \underbrace{\begin{bmatrix} s\mathbf{I}_2 & \mathbf{0} \\ \mathbf{0} & s^{-1}R'_{\frac{\pi}{8}} \end{bmatrix}}_{\text{overlap filtering}} \mathbf{W}_4$$

(3.33)

where s is a scaling factor, $s = 0.8272$, z^{-1} is a delay element, $\mathbf{R}_\theta$ and $\mathbf{R}'_\theta$ are the rotation matrices with an arbitrary rotation with an arbitrary rotation angle θ defined by

$$\mathbf{R}_\theta = \begin{bmatrix} c\theta & s\theta \\ s\theta & -c\theta \end{bmatrix} \triangleq \begin{bmatrix} \cos\theta & \sin\theta \\ \sin\theta & -\cos\theta \end{bmatrix}$$

(3.34)

$$\mathbf{R}'_\theta = \begin{bmatrix} 1 & 0 \\ 0 & 1 \end{bmatrix}\mathbf{R}_\theta \triangleq \begin{bmatrix} c\theta & s\theta \\ -s\theta & c\theta \end{bmatrix}$$

(3.35)

and

$$W_4 = \frac{1}{\sqrt{2}} \begin{bmatrix} I_2 & J_2 \\ J_2 & -I_2 \end{bmatrix} \tag{3.36}$$

$$P_4 = \begin{bmatrix} 1 & 0 & 0 & 0 \\ 0 & 0 & 1 & 0 \\ 0 & 1 & 0 & 0 \\ 0 & 0 & 0 & 1 \end{bmatrix} \tag{3.37}$$

The subscript 4 denotes a 4×4 matrix. I_2 and J_2 are the 2×2 identity and the reversal matrices, respectively. Note that the transform (3.33) is easily constructed from only rotations matrices $R_{\frac{\pi}{4}}$ and $R_{\frac{\pi}{8}}$, scaling factors s, delay element z^{-1}, permutations, and sign inversions. Figure 3.7 shows the lattice structure of the 1D four-channel TDLT.

When a 2×2 input block signal X is two-dimensionally implemented using separable rotation matrices $R\theta_0$ and $R\theta_1$, the output block signal Y is (Suzuki & Yoshida, 2017)

$$Y = R_{\theta_0} X R_{\theta_1}^T \tag{3.38}$$

where T indicates matrix transposition. The 2×2 input block X can be rearranged as $x = [x(0), x(1), x(2), x(3)]^T$. Hence, the output becomes the vector $y = [y(0), y(1), y(2), y(3)]^T$. Here, Equation (3.38) can be expressed as follows:

$$y = R_{\theta_0} \otimes R_{\theta_1} x \tag{3.39}$$

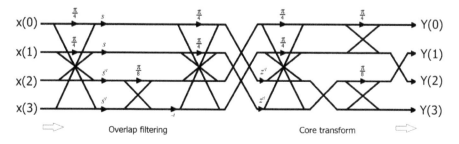

Figure 3.7 Lattice structure of the four-channel hierarchical lapped transform.

where $\otimes$ is the Kronecker product. The 2D non-separable transform matrix $\mathbf{R}_{\theta 0} \otimes \mathbf{R}_{\theta 1}$ can be factorized into multiplierless lifting structures with dyadic lifting coefficients as

$$\mathbf{T}_{\theta o,\theta 1} = \mathbf{R}_{\theta o} \otimes \mathbf{R}_{\theta 1} = \mathbf{W}_4 \cdot \mathrm{diag}\{1,1,-1,-1\}$$

$$\cdot \begin{bmatrix} c\theta_0 - \theta_1 & 0 & s\theta_0 - \theta_1 & 0 \\ 0 & c\theta_0 + \theta_1 & 0 & s\theta_0 + \theta_1 \\ s\theta_0 - \theta_1 & 0 & -c\theta_0 - \theta_1 & 0 \\ 0 & s\theta_0 + \theta_1 & 0 & -c\theta_0 + \theta_1 \end{bmatrix}.$$

$$\cdot \mathbf{W}_4 \cdot \mathrm{diag}\{1,-1,-1,1\} \tag{3.40}$$

The following sets of transforms are defined:

$$\mathbf{T}_{(\pi/4)} \cdot \mathbf{T}_{(\pi/4)} \triangleq \mathbf{T}_{HH} \tag{3.41}$$

$$\mathbf{T}_{(\pi/4)} \cdot \mathbf{T}_{(\pi/8)} \triangleq \mathbf{T}_{HR} \tag{3.42}$$

$$\mathbf{T}_{(\pi/8)} \cdot \mathbf{T}_{(\pi/8)} \triangleq \mathbf{T}_{RR}. \tag{3.43}$$

Suzuki and Yoshida (2017) proposed the following lifting non-separable 2D transforms which result in the best coding performance. The transforms are highly compatible with the JPEG XR standard and that are lower in complexity in terms of the number of operations and lifting steps compared with JPEG XR.

$$\mathbf{T}_{HH} = \mathbf{P}_4 \cdot \begin{bmatrix} 1 & 0 & 0 & 0 \\ -1 & 1 & 0 & 0 \\ -1 & 0 & 1 & 0 \\ -1 & 0 & 0 & 1 \end{bmatrix} \cdot \begin{bmatrix} -1 & 1/2 & 1/2 & 1/2 \\ 0 & 1 & 0 & 0 \\ 0 & 0 & 1 & 0 \\ 0 & 0 & 0 & 1 \end{bmatrix} \cdot \begin{bmatrix} 1 & 0 & 0 & 0 \\ 1 & 1 & 0 & 0 \\ 1 & 0 & 1 & 0 \\ 1 & 0 & 0 & 1 \end{bmatrix}$$
$$\tag{3.44}$$

$$\mathbf{T}_{RR} = \mathbf{P}_4 \cdot \begin{bmatrix} 1 & -3/8 & 0 & 0 \\ 0 & 1 & 0 & 0 \\ 0 & 1 & 1 & 0 \\ 0 & 3/8 & 0 & 1 \end{bmatrix} \cdot \begin{bmatrix} 1 & 0 & 0 & 0 \\ 3/8 & -1 & -7/8 & -3/8 \\ 0 & 0 & 1 & 0 \\ 0 & 0 & 0 & 1 \end{bmatrix} \cdot \begin{bmatrix} 1 & 3/8 & 0 & 0 \\ 0 & 1 & 0 & 0 \\ 0 & -1 & 1 & 0 \\ 0 & -3/8 & 0 & 1 \end{bmatrix}$$
$$\tag{3.45}$$

$$\mathbf{T}_{HR} = \begin{bmatrix} \begin{bmatrix} 1 & 3/8 \\ 0 & 1 \end{bmatrix} \cdot \begin{bmatrix} 1 & 0 \\ -3/8 & 1 \end{bmatrix} & \begin{matrix} 0 & 0 \\ 0 & 0 \end{matrix} \\ \begin{matrix} 0 & 0 \\ 0 & 0 \end{matrix} & \begin{bmatrix} 1 & 3/8 \\ 0 & 1 \end{bmatrix} \cdot \begin{bmatrix} 1 & 0 \\ -3/8 & 1 \end{bmatrix} \end{bmatrix} \cdot [\mathbf{T}_{HH}] \quad (3.46)$$

where

$$\mathbf{R}'_{\pi/8} \approx \begin{bmatrix} 1 & 3/8 \\ 0 & 1 \end{bmatrix} \cdot \begin{bmatrix} 1 & 0 \\ -3/8 & 1 \end{bmatrix}$$

and $\mathbf{P}_4$ is defined in Equation (3.37). Note that $\mathbf{T}_{HH}$ is equivalent to a four-channel Hadamard transform.

3.7 Problems

Pr.3.1.   In Section 3.1, eight types of DCTs are shown. Prove the unitarity property for each type of DCT.

Pr.3.2.   Repeat problem Pr.3.1 for the DSTs shown in Section 3.2.

Pr.3.3.   Develop the shift properties for the DCTs.

Pr.3.4.   Repeat problem Pr.3.3 for the DSTs.

Pr.3.5.   Sketch the waveforms of DCTs, for $N = 4$ and $N = 16$.

Pr.3.6.   Repeat problem Pr.3.5 for the DSTs.

Pr.3.7.   Develop the matrices for all the DCTs for $N = 4$.

Pr.3.8.   Repeat problem Pr.3.7 for the DSTs.

Pr.3.9.   A simple convolution property for DCT-II has been developed by Chitprasert and Rao (1990). The only restriction is that one of the functions to be convolved must be real and even. This property has been utilized in human visual weighted progressive image transmission (Chitprasert & Rao, 1990). Investigate if similar properties exist for the remaining DCTs defined in Equation (3.20).

Pr.3.10.   Repeat Problem 3.9 for the family of DSTs.

4

Entropy Coding

Entropy coding refers to lossless data compression and appears everywhere in modern digital systems. It is a fundamental building block of data compression. Entropy encoders assign a unique prefix-free code to each unique symbol of a source or dictionary. The entropy encoder compresses data by replacing each fixed-length input symbol with the corresponding variable-length prefix-free output code word. The length of each code word is approximately proportional to the negative logarithm of the probability. Therefore, the most common symbols use the shortest codes. Common entropy coders found in compression scheme are the run-length encoding (RLE), Huffman coder, and the Arithmetic coder. In this chapter, the three encoders are explained (Jain, 1989).

4.1 Source Entropy

If $P(E)$ is the probability of an event, its information content $I(E)$. The self-information is expressed as

$$I(E) = \frac{1}{P(E)}. \tag{4.1}$$

The base of the logarithm expresses the unit of information and if the base is 2, the unit is bits. If we have a source with m symbols $\{a_i | i = 1,2,...,m\}$ with probabilities of occurrences $P(a_1), P(a_2),...,P(a_m)$ then the average information per source output for the source z is given by

$$H(\mathbf{z}) = -\sum_{i=1}^{m} P(a_i) \log P(a_i). \tag{4.2}$$

Equation (4.2) is known as the entropy of the source. If the probabilities of the source symbols are known, then the entropy of the source can be measured. For example, if we have the symbols $a_1,...,a_4$ with probabilities

$P(a_1) = 0.2$, $P(a_2) = 0.1$, $P(a_3) = 0.3$, $P(a_4) = 0.4$. The source entropy is given by

$$H(\mathbf{z}) = -0.2\log(0.2) - 0.1\log(0.1) - 0.3\log(0.3) - 0.4\log(0.4) = 1.85 \text{ bits}.$$

The Shannon theorem expresses that *"in any coding scheme, the average code word length of a source of symbols can at best be equal to the source entropy and can never be less than it"* (Shannon, 1948). Therefore, if $m(\mathbf{z})$ is the minimum of the average code word length obtained out of different uniquely decipherable coding schemes, then

$$m(\mathbf{z}) = H(\mathbf{z}). \tag{4.3}$$

The theorem assumes a lossless coding and noiseless coding scheme. The coding efficiency (η) of the coding scheme is given by

$$\eta = \frac{H(\mathbf{z})}{L(\mathbf{z})} \tag{4.4}$$

where $L(\mathbf{z})$ is the average code word length. Since $H(\mathbf{z}) \leq L(\mathbf{z})$ then $0 \leq \eta \leq 1$. The average code word length is given by

$$L(\mathbf{z}) = \sum_{i=1}^{m} L(a_i)P(a_i). \tag{4.5}$$

$L(a_i)$ is the length of the code word assigned to the symbol a_i.

4.2 Run-Length Coder

Data files frequently contain the same character repeated many times in a row. For example, text files use multiple spaces to separate sentences, indent paragraphs, format tables, charts, etc. The RLE runs on sequences having the same value occurring many consecutive times and encodes the length and value of the run. There are a number of RLE variants in common use, which are encountered in the TIFF, PCX, and BMP graphics formats. RLE represents a string by replacing each subsequence of consecutive identical characters with the tuple (char, length). Therefore, the string 11114444222000 would have representation (1, 4) (4, 4) (2, 3) (0, 3). Then compress each tuple by using Huffman or arithmetic coding. This technique works best when the characters repeat often. One such situation is in fax transmission, which contains alternating long sequences of 1's and 0's.

Another variant is to encode the number of repeated characters in a string followed by the value of the next different character (RUNLENGTH,

VALUE). This variant is used in JPEG encoder to encode the AC coefficients of a block of the DCT after quantization, where RUNLENGTH is the number of zeros and VALUE the value of the next coefficient. RLE is terminated by the EOB code (0, 0) (Pennebaker & Mitchell, 1993). Therefore, the string $-102, 4, 0, 10, 60, 3, 5, 0, 0, 0, 0, 2$ is encoded as (0, -102) (0, 4) (1, 10) (0, 60) (0, 3) (0, 5) (4, 2) (0, 0) (Pennebaker & Mitchell, 1993).

4.3 Huffman Coder

Huffman coding is a popular lossless variable length coding (VLC) scheme, based on the following principles (Huffman, 1952):

- Shorter code words are assigned to symbols that are more probable and longer code words are assigned to less probable symbols.
- No code word of a symbol is a prefix of another code word. This makes Huffman coding uniquely decodable.
- Every symbol of the source or dictionary must have a unique code word assigned to it. In JPEG encoder, the Huffman coding is performed on the quantized symbols or after RLE.

The algorithm is based on a binary-tree frequency-sorting method that allows encoding any message sequence into shorter encoded messages and a method to reassemble into the original message without losing any data. The code assignment is based on a series of source reductions. We shall illustrate this with the following example:

1. Arrange the symbols in the decreasing order of their probabilities.

Symbol	Probability $P(a_i)$
a_4	0.4
a_3	0.3
a_1	0.2
a_2	0.1

2. Combine the lowest probability symbols $P(a_1) + P(a_2)$.

Symbol	Probability $P(a_i)$
a_4	0.4
a_3	0.3
$a_1 \vee a_2$	0.3

3. Continue the source reductions of Step 2, until we are left with only two symbols. $P(a_1) + P(a_2) + P(a_3)$

Symbol	Probability $P(a_i)$
a_4	0.4
$a_3 \vee (a_1 \vee a_2)$	0.6

4. Assign codes "0" and "1" to the last two symbols.

Symbol	Probability $P(a_j)$	
a_4	0.4	0
$a_3 \vee (a_1 \vee a_2)$	0.6	1

The compound symbol $a_3 \vee (a_1 \vee a_2)$ was assigned a "1." Therefore, all the elements of this compound symbol will have a prefix "1."

5. Work backwards along the table to assign the codes to the elements of the compound symbols.

Symbol	Probability $P(a_j)$	
a_4	0.4	0
a_3	0.3	10
$a_1 \vee a_2$	0.3	11

Symbol	Probability $P(a_j)$	
a_4	0.4	0
a_3	0.3	10
a_1	0.2	110
a_2	0.1	111

The average code word length is

$$L(\mathbf{z}) = \sum_{i=1}^{m} L(a_i)P(a_i) = 1(0.4) + 2(0.3) + 3(0.2) + 3(0.1) = 1.9 \text{ bits.}$$

The coding efficiency is given by

$$\eta = \frac{H(\mathbf{z})}{L(\mathbf{z})} = \frac{1.85}{1.9} = 0.9737.$$

If we encode the symbols $a_4, a_4, a_4, a_2, a_3, a_3, a_1, a_4, a_4, a_4$, we obtain the encoded bit stream as $\{000\ 11110\ 10\ 11\ 000\}$. We need 15 bits to encode 10 symbols. If we do not use this coding scheme, we need 2 bits per symbol and we would have required 20 bits.

The decoder reverses the coding process. The symbols are recovered perfectly from bit stream with no ambiguity.

1. Examine the leftmost bit in the bit stream.
   a. IF the bit corresponds to one of the symbols add the symbol to the list of decoded symbols and remove the examined bit from the bit stream and go back to step 1.

b. ELSE append the next bit from the left to the already examined bit(s) and examine if the group of bits correspond to the code word of one of the symbols.

   i. IF the group of bits corresponds to one of the symbols add that symbol to the list of decoded symbols, remove the group of examined bits from the bit stream and go back to step 1.

   ii. ELSE go back to step 1.b to append more bits to the group of bits.

There are other variants of this scheme. For example, the source symbols, arranged in order of decreasing probabilities, are divided into a few blocks. Special shift up and/or shift down symbols are used to identify each block and symbols within the block are assigned Huffman codes. This encoding scheme is referred to as Huffman Shift Coding (Firmansah & Setiawan, 2016). The shift symbol is the most probable symbol and is assigned the shortest code word.

4.4 Arithmetic Coder

Arithmetic coding is superior in most respects to the Huffman method (Huffman, 1952). The arithmetic coder encodes the entire message into a single number, a fraction n where $[0.0 \leq n < 1.0)$. Its performance is optimal without the need for blocking of input data. In other words, a message is represented by an interval of real numbers between 0 and 1 (Rissanen & Langdon, 1979), (Witten, Neal, & Cleary, 1987), (Langdon, 1984).

   As the message becomes longer, the interval becomes smaller increasing the number of bits to represent that interval; a number in this interval represents the message and it is known as the tag. Successive symbols of the message reduce the size of the interval according to the symbol probabilities generated by the model. The more likely symbols reduce the interval by less than the unlike alphabet is $\{a_1, a_2, a_3, a_4\}$ and the following fixed model:

Symbol	Probability $P(a_i)$	Range	
a_4	0.4	[0.6, 1.0)	4/10
a_3	0.3	[0.3, 0.6)	3/10
a_2	0.1	[0.2, 0.3)	1/10
a_1	0.2	[0.0, 0.2)	2/10

Assume that the message is a_3, a_2, a_4. Initially, both encoder and decoder know that the range is [0.0, 1.0). After seeing the first symbol a_3, the encoder narrows the range to [0.3, 0.6) to allocate the symbol. The tag lies in this interval. The second symbol, a_2, will narrow this range to one-tenth. This

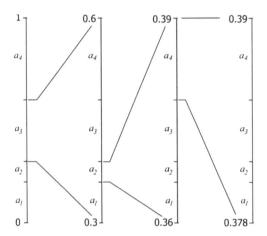

Figure 4.1   Arithmetic coding process with intervals scaled up at each stage.

produces the range [0.36, 0.39) and the tag value is also confined to this interval. Since this range was 0.3 units, one-tenth of this range is 0.03 that corresponds to a_2 which is the second symbol to encode. . Therefore, it is confined from 0.36 to 0.39. The next symbol, a_4, is allocated [0.6, 1.0), which when applied to [0.36, 0.39) gives the smaller range [0.378, 0.39). As more symbols are added to the message, the size of the interval is more reduced.

Figure 4.1 shows the representation of the arithmetic coding. The vertical bars represent the probabilities of the model or the cumulative density function (CDF) or cumulative distribution $F_x(x)$ of the dictionary.

Assume that the decoder knows the encoded range [0.378, 0.39). Hence, it immediately decodes the symbol a_3 from the interval [0.3, 0.6). It is clear that the second symbol is a_2 since it produces the interval [0.36, 0.39) which entirely encloses the interval [0.378, 0.39). The tag is restricted to this interval. Finally, it decodes the last encoded symbol a_4.

We are using decimal numbers to encode the symbols; therefore, the entropy of the three symbols is

$$-\left(\log_{10}(0.3) + \log_{10}(0.1) + \log_{10}(0.4)\right) = -\log_{10}(0.0120) \approx 1.92.$$

It takes about two decimal digits to encode the message. The interval in which a tag resides is disjoint from all intervals for other sequence of symbols. One method for mapping the symbols $\{a_i\}$ to the real numbers $\{i\}$ is using CDF intervals, as

$$\bar{T}_X(a_i) = \sum_{k=1}^{i-1} P(X = k) + \frac{1}{2}P(X = i) = F_X(i-1) + \frac{1}{2}P(X = i). \quad (4.6)$$

$\bar{T}_X(a_i)$ can be used as a unique tag for the symbol a_i. Also, after each observation, we calculate the upper and lower limits of the interval containing the tag of a sequence encoded to that point. The upper and lower limits $u^{(n)}$, $l^{(n)}$ for a sequence $\mathbf{a} = \{a_i \mid i = 1,\cdots,n\}$ of length n can be computed as

$$u^{(n)} = l^{(n-1)} + \left(u^{(n-1)} - l^{(n-1)}\right) F_X\left(a_n\right)$$
$$l^{(n)} = l^{(n-1)} + \left(u^{(n-1)} - l^{(n-1)}\right) F_X\left(a_{n-1}\right). \tag{4.7}$$

Therefore, $\bar{T}_X(\mathbf{a})$ can be computed as the middle point of the interval, in a sequential fashion by using the CDF information from the probability model

$$\bar{T}_X(\mathbf{a}) = \frac{u^{(n)} + l^{(n)}}{2}. \tag{4.8}$$

From the above example, the message is a_3, a_2, a_4, the CDF is $F_X\left(a_n\right) = 0$, $n \le 0$, $F_X\left(a_1\right) = 0.2$, $F_X\left(a_2\right) = 0.3$, $F_X\left(a_3\right) = 0.6$, $F_X\left(a_4\right) = 1$, $F_X\left(a_n\right) = 1$, $n > 4$ and the initial limits are

$$u^{(0)} = 1$$
$$l^{(0)} = 0.$$

The first element of the sequence, a_3, results in the following update:

$$u^{(1)} = 0 + (1-0) F_X(a_3) = 0.6$$
$$l^{(1)} = 0 + (1-0) F_X(a_2) = 0.3.$$

The tag is contained in this interval. The second element of the symbol is a_2

$$u^{(2)} = 0.3 + (0.6 - 0.3) F_X(a_2) = 0.39$$
$$l^{(2)} = 0.3 + (0.6 - 0.3) F_X(a_1) = 0.36.$$

The third symbol is a_4 and the limits of the interval containing the tag are

$$u^{(2)} = 0.36 + (0.39 - 0.36) F_X(a_4) = 0.39$$
$$l^{(2)} = 0.36 + (0.39 - 0.36) F_X(a_3) = 0.378.$$

The tag for the sequence a_3, a_2, a_4 can be generated as

$$\bar{T}_X(\mathbf{a}) = \frac{u^{(3)} + l^{(3)}}{2} = \frac{0.39 + 0.378}{2} = 0.3840.$$

To decipher the tag first, we need to pick an interval $\left[F_x(a_{n-1}), F_x(a_n)\right)$ containing the tag. The only interval containing the tag is the interval [0.3, 0.6) which is the interval of a_3. Therefore, this is the first symbol decoded with lower limit being 0.3 and upper limit being 0.6. We repeat this procedure for the second symbol using the update

$$u^{(2)} = 0.3 + (0.3)F_x(x_n)$$
$$l^{(2)} = 0.3 + (0.3)F_x(x_{n-1}).$$

The only interval that encloses the tag is the interval of a_2; therefore, $x_n = a_2$

$$u^{(2)} = 0.3 + (0.6 - 0.3)F_x(a_2) = 0.39$$
$$l^{(2)} = 0.3 + (0.6 - 0.3)F_x(a_1) = 0.36.$$

For the third symbol, we have

$$u^{(3)} = 0.36 + (0.39 - 0.36)F_x(x_n)$$
$$l^{(3)} = 0.36 + (0.39 - 0.36)F_x(x_{n-1}).$$

The only interval that engulfs the tag is $x_n = a_4$.

Of course, we normally work with binary digits. The method described above is impractical because as we add more symbols to the message, the interval tends to reduce. We want to find a code to represent the sequence **a** in a unique and efficient manner using integer implementations.

4.4.1 Binary Arithmetic Coding (BAC)

Binary arithmetic coding (BAC) is based on the principle of recursive interval subdivision. In the binary arithmetic coding, we use integer implementation (Sayood, 2003). The following example explains the arithmetic coding.

First, we decide a word length of m and map values in the interval [0, 1) to have a total of 2^m binary words. The points 0, 0.5, and 1 of the CDF are mapped to

$$\overbrace{00,\cdots 0}^{m \text{ times}}, \ \overbrace{10,\cdots 0}^{m \text{ times}} \text{ and } \overbrace{11,\cdots 1}^{m \text{ times}},$$

respectively. $F_x(k)$ is computed as

$$F_x(k) = \frac{\displaystyle\sum_{i=1}^{k} n_i}{\text{Total\_Count}} \tag{4.9}$$

with n_r being the number of times the symbols r appears in a sequence of length Total_Count. Then

$$\text{Cum\_Count}(k) = \sum_{i=1}^{k} n_i. \tag{4.10}$$

The intervals for a tag generation of Equation (4.7) can be modified as

$$u^{(n)} = l^{(n-1)} + \left\lfloor \frac{\left(u^{(n-1)} - l^{(n-1)} + 1\right)\text{Cum\_Count}(a_n)}{\text{Total\_Count}} \right\rfloor - 1$$

$$l^{(n)} = l^{(n-1)} + \left\lfloor \frac{\left(u^{(n-1)} - l^{(n-1)} + 1\right)\text{Cum\_Count}(a_{n-1})}{\text{Total\_Count}} \right\rfloor \tag{4.11}$$

where a_n is the nth symbols to be encoded and $\lfloor \bullet \rfloor$ is the floor operation. When the most significant bit (MSB) of both $u^{(n)}$ and $l^{(n)}$ is 1, the tag interval is contained entirely in the upper half of the interval $[00\ldots0, 11\ldots1]$. If their MSB is 0, the tag interval is in the lower half. As the arithmetic is of finite precision, we require the following mapping to rescale the intervals to avoid recording long real numbers:

$$E_1 : [0,0.5) \rightarrow [0,1); \quad E_1(x) = 2x$$
$$E_2 : [0.5,1) \rightarrow [0,1]; \quad E_2(x) = 2(x-0.5)$$
$$E_3 : [0.25,0.75) \rightarrow [0,1); \quad E_3(x) = 2(x-0.25).$$

The principle is: Scale and shift simultaneously x, upper bound, and lower bound will give us the same relative location of the tag. As the intervals narrow, we have one of the following three cases:

1. $\left[l^{(n)}, u^{(n)}\right] \subset [0,0.5) \rightarrow$ output 0 then perform E_1 rescale.
2. $\left[l^{(n)}, u^{(n)}\right] \subset [0.5,1) \rightarrow$ output 1 then perform E_2 rescale.
3. $l^{(n)} \in [0,0.5), \quad u^{(n)} \in [0.5,1) \rightarrow$ output undetermined.

Once we reach case 1 or 2, we can ignore the other half of $[0,1]$ by sending all the prefix bits so far to the decoder. As the arithmetic encoding process progresses, the upper and lower limits, where the tag is enclosed, get closer and closer. To represent these intervals uniquely, we need to increase the precision. In most of the systems, these values are bound to converge because of the limited representation. When the encoder and decoder knows which half the tag is in, we can ignore the other half and we can map the half interval containing the tag to the full range [0, 1) using E1 and E2 to prevent the interval from continually shrinking. During E1 mapping, we send 1, and during E2 mapping, we send 0. Then, we can continue generating another bit of the tag every time the interval is restricted to either half to the unit interval. Note that with this mapping, we lose the information about the significant bit. However, this bit is sent to the decoder.

When the tag interval straddles the middle point of the unit interval (when the lower limit is greater or equal to 0.25 and the upper limit is less than 0.75), we double the tag interval using mapping E3. With E3 mapping, we do not need to send any information to the decoder, but, simply, we keep track of the number of E3 mappings. To see if we need E3 mapping, we observe the second MSB of the upper and lower limits. When this bit for the upper limit is 0 and that for the lower limit is 1, the tag interval lies in the middle half of the interval. Then we complement the second MSB of $u^{(n)}$ and the second MSB of $l^{(n)}$ and shift left shifting 1 in the LSB of $u^{(n)}$ and 0 in the LSB of $l^{(n)}$.

From the previous example, suppose that we want to send the message a_1, a_3, a_2, a_1 with the following counts,

$$\text{Count}(a_1) = 40$$
$$\text{Count}(a_2) = 1$$
$$\text{Count}(a_3) = 9$$
$$\text{Total\_Count} = 50$$

then

$$\text{Cum\_Count}(a_0) = 0$$
$$\text{Cum\_Count}(a_1) = 40$$
$$\text{Cum\_Count}(a_2) = 41$$
$$\text{Cum\_Count}(a_3) = 50$$
$$\text{Scale} = 0.$$

Note that $\text{Count}(a_1) = 40$ and $\text{Count}(a_2) = 41$. We need the value we select for the word length to allow enough range to represent the smallest difference between the endpoints of intervals. Whenever the interval gets small, we rescale. The smallest the interval $\left[l^{(n)}, u^{(n)}\right]$ can be is one-quarter of the total available range of 2^m values. In other words, m should be large enough to accommodate uniquely the set of values between 0 and Total_Count in this case, it is 50. Therefore, the range has to be greater than 200. Hence, $m = 8$ satisfies the requirement. Then,

$$u^{(0)} = 1 = \left(11111111\right) = 255$$
$$l^{(0)} = 0 = \left(00000000\right) = 0.$$

The first symbols to encode is a_1; using Equation (4.11), we obtain

$$l^{(1)} = 0 + \left\lfloor \frac{(255-0+1)\text{Cum\_Count}(a_0)}{50} \right\rfloor = \lfloor 0 \rfloor = 0 = \left(00000000\right)_2,$$

$$u^{(1)} = 0 + \left\lfloor \frac{(255-0+1)\text{Cum\_Count}(a_1)}{50} \right\rfloor - 1 = \lfloor 203.8 \rfloor = 203 = \left(11001011\right)_2.$$

The next symbol is a_3, then

$$l^{(2)} = 0 + \left\lfloor \frac{(203-0+1)\text{Cum\_Count}(a_2)}{50} \right\rfloor = \lfloor 167.28 \rfloor = 167 = \left(10100111\right)_2$$

$$u^{(2)} = 0 + \left\lfloor \frac{(203-0+1)\text{Cum\_Count}(a_3)}{50} \right\rfloor - 1 = 203 = \left(11001011\right)_2.$$

The MSBs are equal. Hence, we shift this value out to the decoder and insert 0 in the LSB of the lower limit and 1 in the LSB of the upper limit

$$l^{(2)} = 78 = \left(01001110\right)_2$$

$$u^{(2)} = 151 = \left(10010111\right)_2.$$

Here, we have a condition for E_3 mapping, when the MSB of the limits are different and the second MSB of the lower limit is 1 and the second MSB of the upper limit is 0. Therefore, we complement the second MSB of both limits and shift 1 bit to the left, shifting in a 0 into the LSB of the lower limit and 1 into the LSB of the upper limit

$$l^{(2)} = 28 = \left(00011100\right)_2$$

$$u^{(2)} = 175 = \left(10101111\right)_2.$$

We also increment Scale to 1. The next symbol is a_2

$$l^{(3)} = 28 + \left\lfloor \frac{(175-28+1)\text{Cum\_Count}(a_1)}{50} \right\rfloor = \lfloor 146.4 \rfloor = 146 = \left(10010010\right)_2$$

$$u^{(3)} = 0 + \left\lfloor \frac{(175 - 28 + 1)\text{Cum\_Count}(a_2)}{50} \right\rfloor - 1 = 148 = (10010100)_2 .$$

The MSBs are equal. Therefore, we shift out 1, shift left by 1 bit

$$l^{(3)} = 36 = (00100100)_2$$

$$u^{(3)} = 41 = (00101001)_2 .$$

As Scale = 1, we transmit 0 and decrement scale by 1. We note that the MSBs are both 0, and we then transmit 0

$$l^{(3)} = 72 = (01001000)_2$$

$$u^{(3)} = 83 = (01010011)_2 .$$

Again the MSBs are 0, and we transmit again 0

$$l^{(3)} = 144 = (10010000)_2$$

$$u^{(3)} = 167 = (10100111)_2 .$$

Both MSBs are 1, and we transmit 1

$$l^{(3)} = 32 = (00100000)_2$$

$$u^{(3)} = 79 = (01001111)_2 .$$

Again the MSBs are 0, and we transmit again 0

$$l^{(3)} = 64 = (01000000)_2$$

$$u^{(3)} = 191 = (10011111)_2 .$$

We increment scale to 1. The next symbol to encode is a_1

$$l^{(4)} = 0 + \left\lfloor \frac{(192)\text{Cum\_Count}(0)}{50} \right\rfloor = \lfloor 0 \rfloor = 0 = (00000000)_2$$

$$u^{(4)} = 0 + \left\lfloor \frac{(192)\text{Cum\_Count}(a_1)}{50} \right\rfloor - 1 = \lfloor 152.6 \rfloor = 152 = (10011000)_2 .$$

The generated bit stream to this point is 1100010. To terminate the process, we send the MSB of the lower limit, and then we see that the scale is 1 so that we send 1 followed by the remaining 7 bits of the lower limit as 1100010010000000.

Scaling solves the problem of maintaining symbol counts because the counts can become arbitrarily large, requiring increased precision arithmetic in the coder and more memory to store the counts themselves. By periodically reducing all symbols' counts by the same factor, we can keep the relative frequencies approximately the same while using only a fixed amount of storage for each count. Scaling allows using lower precision arithmetic (Howard & Vitter, 1992), (Howard & Vitter, 1993).

The following pseudocode is the encoding process explained above:

Initialize $l^{(0)}$, $u^{(0)}$, SCALE, CumCount, TotalCount, $n = 1$
REPEAT
    Get symbol

$$u^{(n)} = l^{(n-1)} + \left\lfloor \frac{\left(u^{(n-1)} - l^{(n-1)} + 1\right)\text{Cum\_Count}(a_n)}{\text{Total\_Count}} \right\rfloor - 1$$

$$l^{(n)} = l^{(n-1)} + \left\lfloor \frac{\left(u^{(n-1)} - l^{(n-1)} + 1\right)\text{Cum\_Count}(a_{n-1})}{\text{Total\_Count}} \right\rfloor$$

WHILE(MSB of $u^{(n)}$ AND MSB of $l^{(n)}$ are equal OR E3 condition holds)
        IF MSB OF $u^{(n)}$ AND MSB OF $l^{(n)}$ ARE EQUAL THEN -- E1 AND E2
                SEND MSB of $u^{(n)}$
                SHIFT $l^{(n)}$ AND $u^{(n)}$ to the left by 1 bit
                INSERT 0 into the LSB OF $l^{(n)}$
                INSERT 1 into the LSB OF $u^{(n)}$
                WHILE SCALE > 0
                        SEND complement of MSB of $u^{(n)}$
                        DEC SCALE
                ENDWHILE
        ENDIF

```
IF E3 THEN
        SHIFT l^{(n)} AND u^{(n)} to the left by 1 bit
        INSERT 0 into the LSB OF l^{(n)}
        INSERT 1 into the LSB OF u^{(n)}
        Complement MSB of l^{(n)} AND u^{(n)}
        INC SCALE
    ENDIF
ENDWHILE
UNTIL no more symbols to encode
```

One strategy to obtain substantial improvements in compression is to use models that are more sophisticated. The increased sophistication invariably takes the form of conditioning the symbol probabilities on contexts consisting of one or more symbols of preceding the current symbol. However, one significant difficulty with using high-order models is that many contexts do not occur often enough to provide reliable symbol probability estimates (Marpe, Schwarz, & Wiegand, 2003).

Other models are adaptive. The simplest adaptive models do not rely on contexts for conditioning probabilities; a symbol's probability is just its relative frequency in the part of the symbols already coded. We shall see that adaptive compression can be improved by taking advantage of locality of reference and especially by using higher order models (Rissanen, 1983), (Rissanen & Langdon, 1979), (Witten, Neal, & Cleary, 1987), (Belyaev, Gilmutdinov, & Turlikov, 2006), (Marpe, Schwarz, & Wiegand, 2003).

4.5 Problems

Pr.4.1.   The entropy (minimum number of bits/symbol for lossless compression) of a source is given by Equation (4.2). Assume that we have $M = 1$ with the following probabilities:

Symbol	Probability
0	0.95
1	0.05

For Huffman code, calculate the entropy per symbol, the average length, and efficiency of the code.

Pr.4.2.   The entropy (minimum number of bits/symbol for lossless compression) of a source is given by Equation (4.2). Assume that we have $M = 2$ with the following probabilities:

Symbol	Probability
00	0.9025
01	0.0475
10	0.0475
11	0.0025

For Huffman code, calculate the entropy per symbol, the average length, and efficiency of the code.

Pr.4.3. A binary image is to be coded in blocks of M pixels. The successive pixels are independent from each other and 5% of the pixels are 1. What are the Huffman codes for $M = 1,2,3$? Also, calculate the compression efficiency of each codebook.

Pr.4.4. A set of images is to be compressed by a lossless method. Each pixel of each image has a value in the range (0–3), i.e., 2 bits/pixel. An image from this set is given below; in this image, the occurrence of pixels of different values is typical of the set of images as a whole.

3	3	3	2
2	3	3	3
3	2	2	2
2	1	1	0

What is compression achievable using the following methods:
a) Huffman coding of the pixel values.
b) Forming differences between adjacent pixels (assuming horizontal raster scan) and then Huffman coding these differences.
c) Run length coding, assuming 2 bits to represent the pixel value and 2 bits to represent the run length.
d) In general, for case a), how many bits does it take to transmit the Huffman codebook from the coder to the decoder? (Assume only that both coder and decoder know that the image has 2 bits/pixel).

Pr.4.5. An image compression algorithm such as JPEG uses "block transforms," what are these? Show that if each block is transformed by a "fast" algorithm such as the fast Fourier transform or fast cosine transform that a 8 x 8 block transform of a 512 x 512 pixel image is almost exactly 3 times faster than transforming the whole image directly.

Pr.4.6. As the block size of problem Pr.4.5 is decreased, the transform becomes faster, what, therefore, sets the limit for the smallest useful block size for image compression? If the correlation coefficient between adjacent pixels is $\rho = 0.9$, why is the choice of 8 for the block size a good one?

Pr.4.7. Why, in general, does the DCT for each block produce a slightly higher degree of compression than using the DFT?

Pr.4.8. Consider an input image given by the two-dimensional function

$$f(x,y) = 2\cos\left[2\pi(4x+6y)\right]$$

which is sampled on an infinite grid of pixels with separation $\Delta x = 0.1$ and $\Delta y = 0.2$, in the x and y directions, respectively. Note that this is less dense than required by Nyquist sampling. Consider that we anyway try to reconstruct the original continuous image from the above sampled version. To do this, we form the continuous FT of the sampled image and multiply by an ideal low-pass spatial filter with cutoff frequencies at half the sampling frequency in each spatial frequency axis and then inverse Fourier transform. Show in your answer what is obtained at each step in the above process, specifically

a) Draw with labeled axes the Fourier transform of the sampled image. Hint 1: make use of the 2D sampling function ("bed of nails") and the convolution theorem. Hint 2: the FT of $\exp\left[-2\pi j(Ax+By)\right]$ is the Diral delta function $\delta[u-A,v-B]$ in the Fourier domain centered on the position $u = A$, $v = B$.

b) Give the Fourier transform after it has been low-pass filter.

c) Show that the reconstructed continuous image is given by the mathematical function $2\cos\left[2\pi(4x+y)\right]$.

d) In order to reconstruct, without distortion, the original image from sampled data, what is the maximum sizes of Δx and Δy that can be used and the size of the low-pass filter?

Pr.4.9. Draw a sketch of the CIE chromaticity diagram with chromaticity coordinates $x = X / (X+Y+Z)$, $y = Y / (X+Y+Z)$. Indicate the following regions:

a) The curved line occupied by the color perceived when single wavelength light from a prism is shone into the eye.

b) The "Line of purples," where purple colors lie.

c) The place where white lies.

d) The line of colors from a blackbody radiating from 3000 to 9000 K.

Pr.4.10. The 1D cosine transform of a length N sequence is defined in the following equation:

$$\left[C_N''\right]_{mn} = \sqrt{\frac{2}{N}}\left[k_n \cos\left(\frac{\pi\left(m+\frac{1}{2}\right)n}{N}\right)\right] \qquad m,n = 0,1,\ldots,N-1. \tag{4.12}$$

Suppose you wish to do a 1D cosine transform in a software package, which only has the capability to do discrete Fourier transforms, carefully describe each of the steps needed to implement it.

Pr.4.11. Suppose we have four symbols a_1, a_2, a_3, a_4. The probabilities and distribution of the symbols are: $P(a_1) = 0.2$, $P(a_2) = 0.5$, $P(a_3) = 0.2$, $P(a_4) = 0.1$. Use arithmetic coding to encode the sequence of symbols $a_3, a_2, a_1, a_1, a_2, a_4$.

Pr.4.12. Suppose we have three symbols a_1, a_2, a_3 and we have a first-order conditional probability model for the source emitting these symbols.

a) For the first symbol in the sequence, we have the probability model: $P(a_1) = 0.2$, $P(a_2) = 0.4$, $P(a_3) = 0.4$.
Encode the sequence a_2, a_1, a_3 using arithmetic coding and plot the intervals as in Figure 4.1.

b) For subsequent symbols in the sequence, we have three context models:

$$\text{Context 1: } P\left(a_1|a_1\right) = 0.1,\ P\left(a_2|a_1\right) = 0.5,\ P\left(a_3|a_1\right) = 0.4$$

$$\text{Context 2: } P\left(a_1|a_2\right) = 0.95,\ P\left(a_2|a_2\right) = 0.01,\ P\left(a_3|a_2\right) = 0.04$$

$$\text{Context 3: } P\left(a_1|a_3\right) = 0.45,\ P\left(a_2|a_3\right) = 0.45,\ P\left(a_3|a_3\right) = 0.1$$

Encode the sequence a_2, a_1, a_3 using context-based arithmetic coding and plot the intervals as in Figure 4.1. Hint: the first symbol is a_2; use $P(a_2) = 0.2$. The second symbol is a_1, and it is after a_2; therefore, use $P\left(a_1|a_2\right) = 0.95$ in context 2. The third symbol is a_3, and it is after a_1; then use $P\left(a_3|a_1\right) = 0.4$ in context 1.
Note: Assume that the decoder knows the probability models.

c) Compare the resulting codes of a) and b).

d) Calculate the entropy per symbol, the average length, and efficiency of the code.

Pr.4.13. Repeat problem Pr.4.11 a) using Huffman code, and calculate the entropy per symbol, the average length, and efficiency of the code and compare with results obtained in Pr.4.11 d).

Pr.4.14.   Using the initial probability model of Pr.4.11 a) and the initial contexts defined in Pr.4.11 to encode the sequence a_2, a_1, a_3, add the symbols a_1, a_1, a_3, a_2, a_2. As you observe the actual frequencies of occurrence, update the probability models to adapt to the actual source.

Note: Because the decoder starts with the same prototype model and sees the same symbols the coder uses to adapt, it can, automatically, synchronize adaptation of its models to the coder.

5

JPEG

Image (and video) coding standards provide interoperability between codecs built by different manufacturers. The standard imposes only restrictions on the bit-stream syntax and decoder. In this way, standards-based products can be built with common software and hardware tools. The encoder is not standardized and its optimization is left to the manufacturer. Standards provide state-of-the-art technology that is developed by a group of experts in the field (Pennebaker & Mitchell, 1993), (Wien, 2015), (Ochoa-Dominguez & Rao, 2019), (Sze, Budagavi, & Sullivan, 2014), (Bing, 2015), (Richardson, 2010), (Taubman & Marcellin, 2001).

JPEG is an image compression standard developed by the *Joint Photographic Experts Group* referred to as the JPEG file interchange format (JFIF). It was formally accepted as an international standard in 1992 (latest version, 1994) as a result of a process that started in 1986. This effort is known as JPEG (JPEG, 2018). Officially, JPEG corresponds to the ISO/IEC international standard 10918-1 (ISO/IEC-10918, 1994).. The text in ISO and ITU-T documents is identical. JPEG is widely used in many application areas and it is still the most dominant still image format around. This file format typically achieves 10:1 compression with little perceptible loss in image quality (Pennebaker & Mitchell, 1993).

5.1 JPEG Parts

JPEG consists of the following seven parts (JPEG, 2018):

Part 1.   Requirements and guidelines: Specifies the core coding system, consisting of the well-known Huffman-coded DCT-based lossy image format, but also including the arithmetic coding option, lossless coding, and hierarchical coding.

Part 2.   Compliance testing: Specifies conformance testing and, as such, provides test procedures and test data to test JPEG encoders and decoders for conformance.

Part 3.   Extensions: Specifies various extensions of the JPEG format, such as spatially variable quantization, tiling, selective refinement, and the still picture interchange file format (SPIFF).

Part 4.   Registration authorities: Registers known application markers, SPIFF tag profiles, compression types, and registration authorities.

Part 5.   File interchange format: Specifies the JPEG file interchange format (JFIF), which includes the chroma up-sampling and YCbCr to RGB transformation.

Part 6.   Application to printing systems: Specifies markers that refine the color space interpretation of JPEG codestreams, such as to enable the embedding of ICC profiles and to allow the encoding in the CMYK color model.

Part 7.   Reference software: Provides JPEG reference software implementations.

JPEG (ISO/IEC-10918, 1994) uses a lossy form of compression based on the discrete cosine transform (DCT). This mathematical operation converts each frame/field of the video source from the spatial (2D) domain into the frequency domain (a.k.a. transform domain). A perceptual model based loosely on the human psycho-visual system discards high-frequency information in the transform domain, i.e., sharp transitions in intensity and color hue.

5.2 Image Pre-Processing

JPEG (ISO/IEC-10918, 1994) introduces a very general image model, able to describe most of the well-known, two-dimensional image representation. Each source image must have a rectangular format and may consist of at least 1 and at most 255 components or planes. Each component may have a different number of pixels in the horizontal than in the vertical axis, but all pixels of all components must be coded with the same number of bits. Each mode defines its own precision. The JPEG algorithm generally does not use the RGB (red, green, and blue) color format. Instead, it uses YCbCr consisting of one luma (Y) component, representing the brightness, and two chroma components (Cb and Cr) as defined by CCIR 601 (256 levels) (ITU-R, 1990). This is desirable as described previously; the human eye tends to notice brightness variation more than color variation. Therefore, we can apply different tolerances and compression ratios to each component to obtain optimal compression at the best image quality possible.

The input components of the image are partitioned into non-overlapping blocks of size 8×8 and the image samples are shifted from unsigned integers with range $\left[0, 2^p - 1\right]$ to signed integers with range $\left[-2^{p-1}, 2^{p-1}\right]$ where p is the number of bits (here, $p = 8$).

Compression is done by breaking a picture down into a number of components called minimum coded units (MCUs). The MCUs are simply made by assembling a number of 8 × 8 pixel sections of the source image to break down the image into workable blocks of data and to allow manipulation of local image correlation at a given part of the image by the encoding algorithm. The number of data units of each component in the MCU is defined in the frame header parameters. When processing each MCU, the algorithm always moves through the component from left to right and then top to bottom order. Besides breaking the image into workable blocks, the MCUs allow manipulation of local image correlation at a given part of the image by the encoding algorithm.

5.3 Encoder

JPEG is a very simple and easy-to-use standard that is based on the discrete cosine transform (DCT). The block diagram of the encoder is shown in Figure 5.1.

The components of the encoder are the following:

a. Image pre-processing: This operation is explained in Section 5.2.
b. DCT or forward discrete cosine transform (FDCT): After image pre-processing, the DCT is applied to each non-overlapping block, each of size 8 × 8 samples. The theory of the DCT has been already discussed in Chapter 3. However, DCT-IDCT pairs recommended in JPEG are Equations (3.27) and (3.28) with the normalization factor $\dfrac{1}{N^2}$ in both equations. The FDCT transforms a block into a set of 64 DCT coefficients. One of these is the DC coefficient and the remaining 63 are the AC coefficients as shown in Figure 5.2. To preserve customization

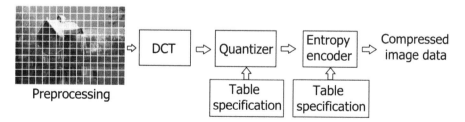

Figure 5.1 General block diagram of the JPEG encoder.

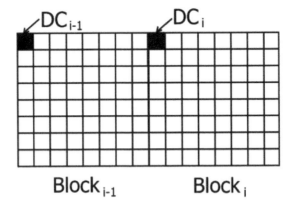

Figure 5.2   Two 8 × 8 transform blocks showing 64 coefficients each. The DC coefficient is black and the remaining 63 AC components are the white squares.

within implementations, JPEG specifies neither any unique FDCT algorithm nor any unique inverse DCT algorithm. The implementations may therefore differ in precision, and JPEG has specified an accuracy test as a part of the compliance test.

c. Quantization: Each of the 64 DCT coefficients of a block is uniformly quantized using a quantization table. The step sizes are chosen as the perceptual threshold or for just noticeable distortion (JND) to compress the image without visible artifacts. Psycho-visual quantization tables appear in ISO-JPEG standard as a recommendation; however, they are not a requirement or mandatory. The uniform quantizer is defined in the following equation:

$$Yq(m,n) = \text{round}\left(\frac{Y(m,n)}{T(m,n)}\right) \qquad (5.1)$$

where ($\bullet$) is the rounding to the nearest integer operation, $Yq(m,n)$ is the quantized coefficients normalized by the quantizer step size, and $T(m,n)$ is the value of the corresponding element from the quantization table. Each table has 64 values that range from 0 to 65,535 (in practice, these values range usually between 0 and 255. Some programmers chose to represent them as 8-bit values; however, to be correct, 16-bit values should be used). A lower number means that less data will be discarded in the compression and a higher quality image should result. Each image has between one and four quantization tables. The most commonly used quantization tables are those published by the Independent JPEG Group (IJG) and shown next for luminance (Table 5.1) and chrominance (Table 5.2).

Table 5.1 Luminance quantization table.

16	11	10	16	24	40	51	61
12	12	14	19	26	58	60	55
14	13	16	24	40	57	69	56
14	17	22	29	51	87	80	62
18	22	37	56	68	109	103	77
24	35	55	64	81	104	113	92
49	64	78	87	103	121	120	101
72	92	95	98	112	100	103	99

Table 5.2 Chrominance quantization table.

17	18	24	47	99	99	99	99
18	21	26	66	99	99	99	99
24	26	56	99	99	99	99	99
47	66	99	99	99	99	99	99
99	99	99	99	99	99	99	99
99	99	99	99	99	99	99	99
99	99	99	99	99	99	99	99
99	99	99	99	99	99	99	99

These tables can be scaled to a quality factor (Q). The quality factor allows the image creation device to choose between larger, higher quality images and smaller, lower quality images. Q can range between 0 and 100 and is used to compute the scaling factor S as follows:

$$S = (Q < 50) ? \frac{5000}{Q} : 200 - 2Q. \tag{5.2}$$

For example, we can scale each entry of IJG standard tables using $Q = 80$ by applying, and $S = 200 - 2(80) = 40$

$$T_s[i] = \left\lfloor \frac{S * T[i] + 50}{100} \right\rfloor \tag{5.3}$$

$$T_s[i] = \left\lfloor \frac{40 * T[i] + 50}{100} \right\rfloor \tag{5.4}$$

where $\lfloor \bullet \rfloor$ is the floor operation.

Tables 5.3 and 5.4 are the scaled laminance and scaled chrominance quantization tables respectively.

    d. Zig-zag scanning: After quantization, the coefficients are zig-zag scanned as shown in Figure 5.3. The zig-zag scanning pattern for run-length coding of the quantized DCT coefficients was established in the original MPEG standard (Ding, Wei, & Chen, 2011). The same pattern is used for luminance and for chrominance

Table 5.3   Scaled luminance quantization table ($Q = 80$).

6	4	4	6	10	16	20	24
5	5	6	8	10	23	24	22
6	5	6	10	16	23	28	22
6	7	9	12	20	35	32	25
7	9	15	22	27	44	41	31
10	14	22	26	32	42	45	37
20	26	31	35	41	48	48	40
29	37	38	39	45	40	41	40

Table 5.4   Scaled chrominance quantization table ($Q = 80$).

7	7	10	19	40	40	40	40
7	8	10	26	40	40	40	40
10	10	22	40	40	40	40	40
19	26	40	40	40	40	40	40
40	40	40	40	40	40	40	40
40	40	40	40	40	40	40	40
40	40	40	40	40	40	40	40
40	40	40	40	40	40	40	40

Note that a value of $Q = 50$ does not affect the values of the quantization tables, T.

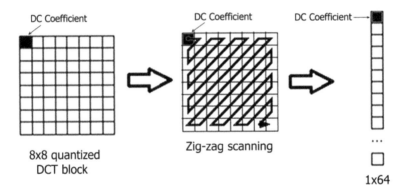

Figure 5.3   Zig-zag order.

e. Entropy encoder: The coefficients are passed to an entropy encoding procedure to further compress the data. The DC coefficients are encoded as a difference from the DC coefficient of the previous block and the current block $\text{DIFF} = DC_i - DC_{i-1}$. The 63 AC coefficients are encoded into (run, level) pair. Afterwards, two alternative variable length (VLC) entropy coding procedures are specified: Huffman and arithmetic coding. If Huffman encoding is used, Huffman table specifications must be provided to the encoder. If arithmetic coding is used, arithmetic coding conditioning table specifications must be provided.

Table 5.5 Coefficient encoding category.

Category	Size	Amplitude range
0	0	0
1	1	−1,1
2	2	−3, −2, 2 ,3
3	3	−7, … , −4, 4, … , 7
4	4	−15, …, −8, 8, …, 15
.	.	.
.	.	.
.	.	.
A	10	−1023, …, −512, 512, … , 1023

Table 5.6 Suggested Huffman code for luminance DC differences.

Category	Code length	Code word
0	2	00
1	3	010
2	3	011
3	3	100
4	3	101
5	3	110
6	4	1110
.	.	.
.	.	.
.	.	.
10	8	11111110
11	9	111111110

Coding of DC coefficients: DC coefficients may vary greatly over the whole image but slowly from one block to its neighbor (once again, zig-zag order). Therefore, to remove the correlation between DC coefficients of neighbor blocks, JPEG applies the differential pulse code modulation (DPCM). The DPCM-coded DC coefficients are represented by a pair of symbols (SIZE, AMPLITUDE), where SIZE is the number of bits to represent a coefficient and AMPLITUDE is the actual number of bits. DPCM may require more than 8 bits and might be negative. One's complement is used for negative numbers. Table 5.5 shows 10 out of the 16 coefficient encoding categories.

Table 5.6 shows the suggested Huffman code for luminance DC differences.

For example, if the first five luminance DC coefficients are 150, 155, 149, 152, and 144, corresponding to the first five blocks, the outputs of the DPCM are 150, 5, −6, 3, and −8.

The codes become (9, 010010110), (4, 0101), (4, 1001), (3, 0111), and (5, 10111). Then, the SIZE is Huffman coded (expect many small SIZEs). The

Table 5.7   Suggested Huffman code for chrominance DC difference.

Category	Code length	Code word
0	2	00
1	2	01
2	2	10
3	3	110
4	4	1110
.	.	.
.	.	.
.	.	.
10	10	1111111110
11	11	11111111110

AMPLITUDE is not Huffman coded (uniform distribution is expected and without much coding gain).

For example, the number -8 is coded as 101 0111 where 0111 is the one's complement of 8 and 101 the size from Table 5.6. Table 5.7 shows the suggested Huffman code for chrominance DC difference.

Coding of AC coefficients: After quantization, many AC coefficients become zero. Therefore, run-length coder (RLC) is applied to the 63 AC coefficients. Now, the RLC step replaces values by a pair (RUNLENGTH, VALUE), where RUNLENGTH is the number of zeros in the run and VALUE is the next non-zero value. VALUE is really a (SIZE, AMPLITUDE) pair. Therefore, RUNLENGTH is the number of consecutive zero-lined AC coefficients preceding the non-zero AC coefficient. The value of RUNLENGTH is in the range 0–15, which requires 4 bits for its representation.

SIZE is the number of bits used to encode AMPLITUDE. The number of bits for AMPLITUDE is in the range of 0–10 bits; so 4 bits are needed to code SIZE.

AMPLITUDE is the amplitude of the non-zero AC coefficient in the range of [+1024 to −1023], which requires 10 bits for its coding.

For example, suppose that the following combination of quantized coefficients resulted after the zig-zag scanning, including the DC coefficient $(32, 6, -1, -1, 0, -1, 0, 0, 0, -1, 0, 0, 1, 0, 0, ..., 0)$.

The RLC becomes $(0, 6), (0, -1), (1, -1), (3, -1), (2, 1), (0, 0)$, note that the DC coefficient is ignored and $(0, 0)$ is end of block (EOB). RUNLENGTH and SIZE are each 4-bit values stored in a single byte – *Symbol1*. For runs greater than 15, special code $(15, 0)$ is used. *Symbol2* is the AMPLITUDE. *Symbol1* is entropy coded but *Symbol 2* is not. The pair of symbols represents each AC coefficient as *Symbol-1*: (RUNLENGTH, SIZE); *Symbol-2*: (AMPLITUDE).

Symbol-1 is VLC coded using a Huffman table shown on the right: Up to two separate custom Huffman tables can be specified within the image header for AC coefficients. *Symbol-2* is coded in sign-magnitude format as for DC coefficients.

Entropy coding is applied to the RLC coded AC coefficients. The baseline entropy coding method uses Huffman coding on images with 8-bit components. Table 5.8 shows the partial Huffman for luminance for AC Run/Size pairs and Table 5.9 for chrominance. ZRL is the special symbol representing 16 zeros.

For example, the symbol representation of the number 476 in the sequence 0,0,0,0,0,0,476 is (6, 9) (476), where RUNLENGTH = 6, SIZE = 9, and AMPLITUDE = 476. If RUNLENGTH is greater than 15, then *Symbol-1* (15,0) is interpreted as the extension symbol with RUNLENGTH = 16. These can be up to three consecutive (15, 0) extensions, i.e., (15, 0) (15, 0) (7, 4) (12) RUNLENGTH is equal to 16 + 16 + 7 = 39, SIZE = 4, and AMPLITUDE = 12..

The second step in Huffman coding is converting the intermediate symbol sequence into binary sequence. In this phase, symbols are replaced with variable length codes, beginning with the DC coefficients, and continuing

Table 5.8 Partial Huffman table for luminance for AC Run/Size pairs.

Run/ Size	Code length	Code word
0/0 (EOB)	4	1010
0/1	2	00
0/2	2	01
0/3	3	100
0/4	4	1011
0/5	5	11010
0/6	7	1111000
0/7	8	11111000
0/8	10	1111110110
0/9	16	1111111110000010
0/A	16	1111111110000011
1/1	4	1100
1/2	5	11011
1/3	7	1111001
1/4	9	111110110
1/5	11	11111110110
1/6	16	1111111110000100
...	...	...
1/A	16	1111111110001000
...	...	...
15/0 (ZRL)	11	11111111001
... 15/A	...	...
	16	1111111111111110

Table 5.9   Partial Huffman table for chrominance for AC Run/Size pairs.

Run/ Size	Code length	Code word
0/0 (EOB)	2	00
0/1	2	01
0/2	3	100
0/3	4	1010
0/4	5	11000
0/5	5	11001
0/6	6	111000
...	...	...
0/A	12	111111110100
1/1	4	1011
1/2	6	111001
1/3	8	11110110
...	...	...
15/0 (ZRL)	10	1111111010
15/1	15	111111111000011
15/2	16	1111111111110110
...	...	...
15/A	16	1111111111111110

with AC coefficients. Each *Symbol-1* is encoded with a VLC, obtained from the Huffman table set specified for each image component. The generation of Huffman tables is discussed by Pennenbaker and Mitchell (1993). *Symbol-2* is encoded using a variable-length integer (VLI) code, whose length in bits is given in Table 5.5.

For example, for an AC coefficient presented as the symbols (1, 4) (12) the corresponding binary presentation will be: (111110110 1100), where (111110110) is VLC obtained from the Huffman table, and (1100) is the VLI code for 12.

Figure 5.4 shows the Huffman entropy encoder (Pennebaker & Mitchell, 1993). The encoding process is as follows:

1. The statistical model translates the descriptors into symbols. Each symbol is assigned to a single code word or probability.
2. The adaptor assigns code words in case of Huffman coding, or probability estimates, in case of arithmetic coding, needed by the encoder.
3. Code table is stored in memory and contains the Huffman code table or the arithmetic coding probability estimates. Therefore, the outputs of the code table module are the code words or probability estimates that are sent to the encoder section.
4. The encoder takes the code words to probability estimates generated by the code table and converts the symbols to bits.

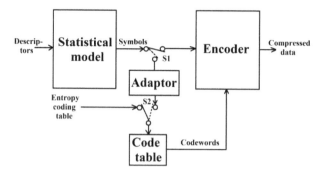

Figure 5.4 Huffman entropy encoder.

Arithmetic coding is a part of the JPEG standard. Initially, encoders and decoders due to patents did not implement it. Currently, there is no such limitation. JPEG uses an adaptive binary arithmetic coder. The descriptors are translated into binary decisions by a statistical model, which also generates a context. The context selects a particular probability estimate.

Figure 5.5 shows the four basic building blocks of the arithmetic entropy encoder; note that the adaptor is not optional. Therefore, the arithmetic encoder uses an adaptive binary scheme. The coding process is as follows (Pennebaker & Mitchell, 1993):

1. The statistical model translates the descriptors and generates a context that selects a particular probability estimate to be used in coding the binary decision. The context is sent to the adaptor and the binary symbols to the encoder.
2. The arithmetic coding conditioning table provides parameters in generating the contexts.
3. The statistics area generates a probability estimate based on the context. The probability estimate is sent to the encoder.

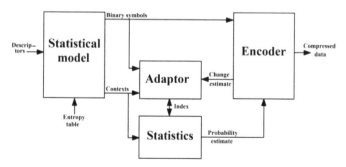

Figure 5.5 Arithmetic entropy encoder.

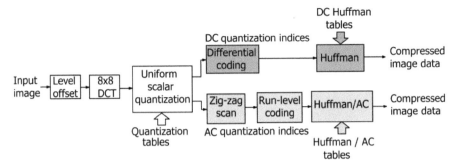

Figure 5.6   JPEG encoder block diagram.

4. The encoder encodes the symbol based on the probability estimate supplied by the statistics area. The encoder also signals the adaptor to make the probability estimate larger or smaller for a particular context.

Figure 5.6 shows a complete block diagram of the JPEG encoder explained above.

5.4 Modes of Operation

The JPEG standard is applicable to both gray-scale and color images. Compression is carried out by splitting a picture into a number of components. For example, a gray-scale image has only one component and a color image usually has three components. Each component has x columns and y rows of samples which are simply the pixel data of the image. In JPEG baseline, the number of samples is the same as the resolution of the image. A single pass through one component and its samples is known as a scan. A scan contains a complete encoding of one (non-interleaved) or more image components (interleaved). For example, using YCbCr format with 4:2:0 sampling, there are three components. If the compressed data is non-interleaved, there would be only three scans. If the compressed data is interleaved, there would be only one scan.

Hence, there are two methods of processing MCUs and components. The first method is whereby each MCU contains data from a single component only and, hence, is composed of only one unit of data. The second method is where the data from each component is interleaved within a single MCU. This means that each MCU contains all the data for a particular physical section of an image, rather than a number of MCUs containing each component of that section. If the compressed image data is non-interleaved, the MCU is defined to be one data unit (8 × 8 block for DCT-based processes). If the compressed data is interleaved, the MCU contains one or more data units from each component.

The number of JPEG modes of operation is four:

1. Baseline or sequential DCT-based mode: The image components are compressed either individually or in groups by interleaving. The interleaving allows compressing and decompressing color images with a minimum of buffering. Sequential mode supports sample data with 8 and 12 bits of precision. In the sequential JPEG, each color component is completely encoded in single scan. An 8 × 8 forward DCT is applied to each block, and the resulting coefficients are quantized and entropy coded. Within the sequential mode, two alternate entropy encoding processes are defined: one uses Huffman coding and the other uses arithmetic coding. Figure 5.7 shows the block diagram of the sequential JPEG encoder and decoder.

   In baseline mode, JPEG starts to display the image as the data is made available line-by-line as shown in Figure 5.8. This JPEG format is recognizable to most web browsers.

The image components are compressed individually or in groups, a single scan coding a component or group of components in one pass.

The decoders require the following:

- DCT-based process.
- 8-bit per samples.
- Huffman coding (2 AC and 2 DC tables).
- It shall process scans with 1, 2, 3, and 4 components.
- Interleaved and non-interleaved scans.
- Sequential decoding.

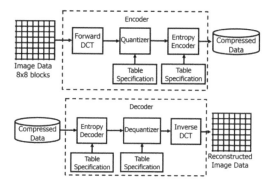

Figure 5.7   Block diagram of sequential JPEG encoder and decoder.

Start ——————————————→ End

Figure 5.8   Display of an image using JPEG baseline.

2. Progressive DCT-based, also known as extended DCT-based: Similar to the sequential DCT-based mode, in this mode, each block is DCT transformed and quantized. The quantized coefficients data are encoded by a sequence of scans that can be performed in two ways:

a. Spectral selection: Coefficients are grouped into spectral bands of related spatial frequency as shown in Figure 5.9(a). The lower-frequency bands are sent first to the decoder. In spectral selection, DCT coefficients are re-ordered as zig-zag sequence and the band containing the DC coefficient is encoded at the first scan as shown in Figure 5.9(b).

b. Successive approximation: Data is first sent with lower precision and then refined. Successive approximation yields better quality at lower bit rates. In this approximation, significant bits of DCT coefficient are encoded in the first scan. Each succeeding scan improves the precision of the coefficients by one bit until full precision is reached.

c. Mixture: Combines the spectral selection and successive approximation.

The compressed data for each component is placed in a minimum of 2 and as many as 896 scans. The initial scans create a rough version of the image,

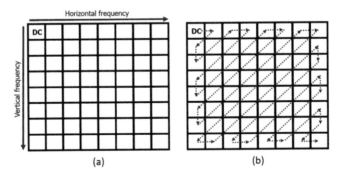

Figure 5.9   Block of 8 × 8 DCT coefficients. (a) Spectral bands. (b) Zig-zag scan.

Start ⟶ End

Figure 5.10 Progressive JPEG.

while subsequent scans refine it. Figure 5.10 shows an image recovered using progressive JPEG. Note how the resolution is improved each time.

The decoders require the following:
- DCT-based process.
- 8 bits or 12 bits per samples.
- Huffman or arithmetic coding (4 AC and 4 DC tables).
- It shall process scans with 1, 2, 3, and 4 components.
- Interleaved and non-interleaved scans.
- Sequential or progressive decoding.

3. Sequential DPCM predictive coding based (lossless process): Preserves exactly the original image with smaller compression ratio. In this mode, the reconstructed neighbor samples (a, b, and c) are used to predict the current sample (x) using Table 5.10. Figure 5.11 shows the neighboring samples necessary to predict the current sample.

Predictor 1 of Table 5.10 is used to encode the first line of an image component because this line is always encoded in 1D.

The decoders require the following:
- Predictive process (not DCT-based process).
- 2–16 bits per samples.
- Huffman or arithmetic coding (4 DC tables).

Table 5.10 Predictors for lossless coding for a given scan though a component or group of components.

Selection	Prediction
0	No prediction
1	$x = a$
2	$x = b$
3	$x = c$
4	$x = a + b - c$
5	$x = a + (b - c)/2$
6	$x = b + (a - c)/2$
7	$x = (a + b)/2$

c	b
a	x

Figure 5.11.   Neighboring samples (*a*, *b*, and *c* = reconstructed samples; *x* = sample under prediction).

- It shall process scans with 1, 2, 3, and 4 components.
- Interleaved and non-interleaved scans.
- Sequential decoding.
4. Hierarchical mode: The main feature of this mode is the encoding of the image at different resolutions as shown in Figure 5.13. This mode uses extensions of either DCT or predictive-based modes. This mode provides for progressive coding with increasing spatial resolution between progressive stages. In this mode, the image is broken down into a number of subimages called frames as shown in Figure 5.12. A frame is a collection of one or more scans.

The first frame creates a low-resolution version of image. Hierarchical coding provides a progressive presentation similar to progressive DCT-based mode but is useful in applications that have multiresolution requirements and provides the capability of progressive coding to a final lossless stage.

The decoders require the following:

- Multiple frames (differential and non-differential processes).
- Extended-based or lossless process.

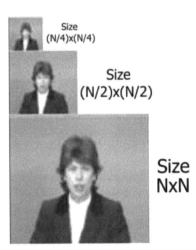

Figure 5.12   Hierarchical images resulting after hierarchical mode.

- It shall process scans with 1, 2, 3, and 4 components.
- Interleaved and non-interleaved scans.

The encoding procedure can be summarized as follows:
1. Filter and down-sample the original image by factors of 2 horizontally and/or vertically.
2. Encode the down-sampled image using a sequential DCT, progressive DCT, or lossless encoder.
3. Decode the down-sampled image and interpolate it (filter + up-sample) by a factors of 2 horizontally and/or vertically.
4. The interpolated (filter + up-sample) image can be used as a prediction of the original image at this resolution and encode the difference image using one of the sequential DCTs, progressive DCT, or lossless encoders described previously.
5. Repeat steps 3) and 4) until the full resolution of the image has been encoded.

Hierarchical encoding is useful in applications to access high-resolution images with a lower-resolution display.

Figure 5.13 shows a three-level hierarchical encoder; the image is low-pass filtered and subsampled by 4:1, in both directions, to give a reduced image size 1/16. The baseline encoder then encodes the reduced image. The decoded image at the receiver may be interpolated by 1:4 to give the full size image

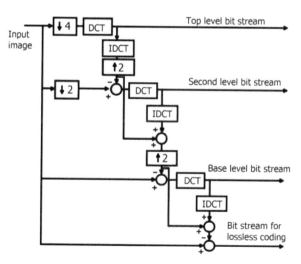

Figure 5.13 A three-level hierarchical encoder.

for display. At the encoder, another baseline encoder encodes the difference between the subsampled input image by 2:1 and the 1:2 up-sampled decoded image. By repeating this process, the image is progressively coded, and at the decoder, it is progressively built up. The bit rate at each level depends on the quantization step size at that level. Finally, for lossless reversibility of the coded image, the difference between the input image and the latest decoded image is lossless entropy coded (no quantization). The hierarchical mode offers a progressive representation similar to the progressive DCT-based mode and offers the capability of progressive transmission to a final lossless stage.

JPEG divides the compression sequence into frames and scans. For non-hierarchical encoding, the frame defines the basic attributes of the image and only a single frame is allowed, while in the hierarchical mode, a sequence of frames is allowed. For each component in a frame, the encoder inserts a component identifier (horizontal and vertical sampling factors) and specifies the quantization table. The number of components in the scan indicates the type of data ordering used in the scan, i.e., with one component non-interleaved data, each MCU contains only one data unit. If the scan contains more than one component, the data are interleaved.

5.5 Some Limitations of JPEG Encoder

- Low bit-rate compression: JPEG offers an excellent quality at high and mid bit rates. However, the quality is unacceptable at low bit rates (e.g., below 0.25 bpp for 8 bpp image).
- Lossless and lossy compression: JPEG cannot provide a superior performance at lossless and lossy compression in a single codestream.
- Transmission in noisy environments: The current JPEG standard provides some resynchronization markers, but the quality still degrades when bit errors are encountered.
- Different types of still images: JPEG was optimized for natural images. Its performance on computer generated images and bi-level (text) images is poor.
- Iterating compression over multiple processing stations (i.e., an in-house intranet cabling) piles up latency and introduces iterative re-compression with additional compression defects.
- JPEG does not guarantee a maximal latency of 32 lines due to its variable rate Huffman entropy coding.
- JPEG is less suitable than JBIG for working with text or monochrome graphics with clear boundaries.

5.6 Table of Metadata Structure

A JPEG file contains several segments containing different types of data, delimited by 2-byte codes or hexadecimal markers (some markers are just 1-byte). The markers begin with 0xFF and end with a code (1 byte) that indicates the type of marker. Some markers consist of just those 2 bytes; others are followed by 2 bytes indicating the length of marker-specific payload data that follows. The length includes the 2 bytes for the length but not the 2 bytes for the marker. The metadata in JPEG file is stored in APPn (0xFF, 0xEn) segment and the comment is stored in COM segment (0xFF, 0xFE). Several vendors might use the same APPn marker type to include their information; in case of JPEG XT (JPEG, 2018), this marker is called APP11. Some markers often begin with a vendor name (e.g., "Exif" or "Adobe") or some other identifying string. Exiv2 provides fast and easy read–write access to the Exif, IPTC, and XMP. The common JPEG markers are shown in Table 5.11.

Table 5.11   The common JPEG markers.

Short name	Bytes	Payload	Name and comments
SOI	0xFF, 0xD8	None	Start of Image
SOF0	0xFF, 0xC0	Variable size	Start of Frame (baseline DCT) – Huffman Indicates that this is a baseline DCT-based JPEG and specifies the width, height, number of components, and component subsampling
SOF1	0xFF, 0xC1	Variable size	Start of Frame (extended sequential DCT) – Huffman
SOF2	0xFF, 0xC2	Variable size	Start of Frame (progressive DCT) Indicates that this is a progressive DCT-based JPEG and specifies the width, height, number of components, and component subsampling – Huffman
SOF3	0xFF, 0xC3	Variable size	Start of Frame (lossless sequential) – Huffman
SOF5	0xFF, 0XC5	Variable size	Differential sequential DCT – Huffman
SOF6	0xFF, 0XC6	Variable size	Differential progressive DCT – Huffman
SOF7	0xFF, 0XC7	Variable size	Differential lossless DCT – Huffman
SOF9	0xFF, 0XC9	Variable size	Extended sequential DCT – Arithmetic
SOF10	0xFF, 0XCA	Variable size	Progressive DCT – Arithmetic
SOF11	0xFF, 0XCB	Variable size	Lossless (sequential) – Arithmetic
SOF13	0xFF, 0xCD	Variable size	Differential sequential DCT – Arithmetic
SOF14	0xFF, 0Xce	Variable size	Differential progressive DCT – Arithmetic
SOF15	0xFF, 0XCF	Variable size	Differential lossless DCT – Arithmetic
DHT	0xFF, 0xC4	Variable size	Define Huffman Table(s)
DQT	0xFF, 0xDB	Variable size	Define Quantization Table(s)

Table 5.11   *Continued*

Short name	Bytes	Payload	Name and comments
DRI	0xFF, 0xDD	2 bytes	Define Restart Interval Specifies the interval between RSTn markers in macroblocks. This marker is followed by 2 bytes indicating the fixed size so that it can be treated like any other variable size segment.
DNL	0xFF, 0xDC	4 bytes	Define restart interval
SOS	0xFF, 0xDA	Variable size	Start Of Scan Begins a top-to-bottom scan of the image. In baseline DCT JPEG images, there is generally a single scan. Progressive DCT JPEG images usually contain multiple scans. This marker specifies which slice of data it will contain and is immediately followed by entropy-coded data.
RSTn	0xFF, 0xDn $n(n = 0...7)$	None	Restart Inserted every r macroblocks, where r is the restart interval set by a DRI marker. Not used if there was no DRI marker. The low 3 bits of the marker code cycle in value from 0 to 7.
APPn	0xFF, 0xEn	Variable size	Application-specific
COM	0xFF, 0xFE	Variable size	Comment
DAC	0xFF, 0xCC	Variable size	Define arithmetic coding tables (s)
DHP	0xFF, 0xDE	Variable size	Define hierarchical progression
EOI	0xFF, 0xD9	None	End Of Image

Table 5.12   Example of segments of an image coded using baseline DCT.

Type of segment	Length of segment
Start of image	0
APP0	16
Quantization table	67
Start of frame: baseline DCT	11
Huffman table	28
Huffman table	63
Start of scan	49,363
End of image	0

APP1 segment stores a great amount of information on photographic parameters for digital cameras and it is the preferred way to store thumbnail images nowadays. It can also host an additional section with GPS data. Table 5.12 is an example of segments that can be found in the codestream of an image.

The APP0 segment marks this file as a JFIF/JPEG. JFIF defines some extra fields (like the image resolution and an optional thumbnail).

6

JPEG XR

JPEG-eXtended Range (JPEG XR) corresponds to the (ITU-T Rec. T.832|ISO/ IEC 29199-2) international standard. JPEG XR is a lossy and lossless image compression and file format developed by Microsoft as HD Photo (formerly Windows Media Photo) (Microsoft, 2018). This standard is an image coding system primarily targeting the representation of continuous-tone still images such as photographic images. It is designed to address the limitations of today's formats and to simultaneously achieve high image quality while limiting computational resource and storage capacity requirements. Moreover, it aims at providing a new set of useful image coding capabilities focused around high/extended dynamic range imagery. The "XR" part of the JPEG XR name evokes the intent of the design to apply to an "extended range" of applications beyond the capabilities of the original baseline JPEG standard. JPEG XR is based on the "HD Photo" technology developed by Microsoft to address the demands of consumer and professional digital photography. It is expected that JPEG XR will help pioneer inventive and groundbreaking products and services in the marketplace and will bring widespread added value for consumers and professionals alike (Doufaux, Sullivan, & Ebrahimi, 2009).

JPEG XR (or its compatible HD Photo predecessor) is supported in the Microsoft Windows Vista and Windows 7 operating systems and in some configurations of Windows XP. This includes support of various features such as image viewing in Windows Photo Gallery and Windows Live Photo Gallery, thumbnail preview in the Windows Explorer shell, and image management in the Windows Imaging Component (WIC) system. Various other Microsoft-led efforts have included support for the feature, such as the SeaDragon, Photosynth, Image Composite Editor, and HD View initiatives (Doufaux, Sullivan, & Ebrahimi, 2009).

JPEG (JPEG, Overview of JPEG, 2018) uses a bit depth of 8 for each of three channels, resulting in 256 representable values per channel (a total of 16,777,216 representable color values) (JPEG, Overview of JPEG XR, 2018). The original JPEG specifications do include a 12-bit mode. The lossless

JPEG coding mode supports up to 16 bits per sample. However, these two formats are incompatible with the 8-bit mode and rarely are used in practical applications (Artusi *et al.*, 2016).

More demanding applications may require a bit depth of 16, providing 65,536 representable values for each channel, and resulting in over 2.8×10^{14} color values. Additional scenarios may necessitate even greater bit depths and sample representation formats. When memory or processing power is at a premium, as few as five or six bits per channel may be used. 16-bit and 32-bit fixed point color component codings are also supported in JPEG XR. In such encodings, the most-significant 4 bits of each color channel are treated as providing additional "headroom" and "toe room" beyond the range of values that represents the nominal black-to-white signal range. However, 16-bit and 32-bit floating-point color component coding are supported in JPEG XR. Besides RGB and CMYK formats, JPEG XR also supports gray-scale and multichannel color encodings with an arbitrary number of channels.

6.1 JPEG XR Parts

JPEG XR (JPEG, Overview of JPEG XR, 2018) is a multi-part specification and it currently includes the following parts:

Part 1.   System architecture: The document is an overview of the different parts of the specifications and gives some guidelines on the best encoding/decoding practices. This part is a non-normative technical report (TR). The overview of JPEG XR coding technology includes a description of the supported image formats, the internal data processing hierarchy and data structures, the image tiling design supporting hard and soft tiling of images, the lapped bi-orthogonal transform, supported quantization modes, adaptive coding and scanning of coefficients, entropy coding, and, finally, the codestream structure.

Part 2.   Image coding specification: This part specifies the image coding format.

Part 3.   Motion JPEG XR: This part specifies the use of JPEG XR encoding for stored sequences of moving images with associated timing information. The Motion JPEG XR file format is based on the ISO Base Media File Format standard (ISO BMFF) (ISO/IEC-14496-12, 2015).

Part 4.   Conformance testing: This part specifies a set of tests designed to verify whether codestreams, files, encoders, and decoders meet the normative requirements specified in Part 2. The tests specified provide methods to (non-exhaustively) verify whether encoders and decoders meet these requirements.

Part 5. Reference software: The reference software can aid adoption of standard by providing an example implementation that can be used as a basis for making encoder and decoder products and can be used to test conformance and interoperability as well as to demonstrate the capabilities of the associated standard. It includes both encoder and decoder functionality.

JPEG XR uses a frequency decomposition scheme that is a hybrid between the JPEG and JPEG 2000 approaches (Faria, Fonseca, & Costa, 2012). Like the original JPEG standard, its frequency decomposition uses block-based transformations. However, an overlapping processing stage is employed to lengthen the basis functions and to enhance theoretical coding gain beyond what would be provided by an ordinary block transform of the same basic block size. Additionally, the block transform is re-applied in a hierarchical fashion, which gives its subband decomposition some substantial similarities to that of a wavelet transform. Figure 6.1 shows the block diagram of the JPEG XR (a) encoder and (b) decoder (Faria, Fonseca, & Costa, 2012), (ITU-T-Rec-T.832|ISO/IEC-29199-2, 2009).

Note that the JPPEG XR is a block-based image coder similar to the traditional image coding paradigm and that includes transformation and coefficient-encoding stages. JPEG-XR employs a reversible integer-to-integer

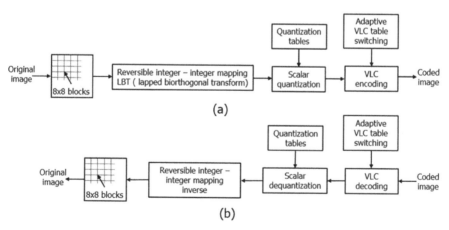

Figure 6.1 Block diagram of JPEG XR (a) encoder and (b) decoder (ITU-T-Rec-T.832 | ISO/IEC-29199-2, 2009).

mapping, called lapped biorthogonal transform (LBT) to map the input blocks from the spatial domain to the frequency domain.

6.2 Image Pre-Processing

Both JPEG 2000 and JPEG XR use the concept of "tiling" which partitions the input image in rectangular regions. Each tile is independently encoded. This subdivision allows to specify a context for image coding or areas of the image called "regions of interest" to be manipulated by the system-level architecture (see Figure 6.2).

6.2.1 Pre-Scaling

JPEG XR provides a much more flexible approach to the numerical encoding of image data by supporting a wide range of different pixel formats. JPEG XR supports three basic types of numerical encoding, each at a variety of bit depths – specifically, these types are unsigned integer, fixed-point, and floating-point representations. A variety of pixel formats is supported by this format using unsigned integer representations in bit depths of 8 and 16 as well as smaller bit depths for specialized applications. JPEG XR also supports a number of additional pixel formats that avoid some problematic aspects involved in the unsigned integer representations.

Pre-scaling is applied for 16-bit unsigned integer (BD16), 16-bit signed integer (BD16S), 16-bit float (BD16F), 32-bit signed integer (BD32S), 32-bit float (BD32F), and RGBE. They are usually used when the input data range is greater than 27/24 bits. The 27-bit limit is used when data is scaled, and the 24-bit limit applies when the data is unscaled. For the most common cases such as for 16-bit data, the pre-scaling steps are omitted although they may still be used.

For OUTPUT_BITDEPTH equal to BD16, BD16S, or BD32S, input values are right-shifted by the value specified by SHIFT_BITS. For OUTPUT_BITDEPTH equal to BD16F, a sign bit extension is applied. For OUTPUT_BITDEPTH equal to BD32F, the encoder first selects a value of mantissa length LEN_MANTISSA, and an exponent bias EXP_BIAS. LEN_MANTISSA is less than or equal to 23 (Rec. ITU-T, 2016), (ISO/IEC-29199, 2010).

6.2.2 Color Conversion

JPEG XR provides native support for both RGB and CMYK color types by converting these color formats to an internal luma-dominant format through the use of a reversible color transform. In addition, YUV, monochrome, and arbitrary *n*-channel color formats are supported.

The encoder uses a reversible color format conversion to convert between the output color format OUTPUT_CLR_FMT and internal color format INTERNAL_CLR_FMT. In order to convert from RGB to YUV422 or YUV420, down-sampling must be performed after color format conversion. RGB to YONLY conversions is obtained by discarding the U and V components on the encoder side 23 (Rec. ITU-T, 2016), (ISO/IEC-29199, 2010).

Prior to color conversion, a bias is subtracted from all values to zero center their range. The amount of the bias is determined by the source bit depth. When the scaling mode is used, the color values are shifted left prior to encoder color conversion.

The input image is composed of a primary image plane and, when present, an alpha image plane. The primary plane is formed of components. The number of components can be NumComponents = 1 and up to NumComponents = 4111. Each component is an array of ExtendedHeight[i] × ExtendedWidth[i] samples, where ExtendedWidth[i] and ExtendedHeight[i] specify (respectively) the width and height of the array for the *i*th component, and $0 <= i <$ NumComponents. The index $i = 0$ refers to the luma component of the primary or the alpha image plane. The alpha image plane contains only one component with same dimensions as those of the luma component of the primary image plane.

The alpha image plane indicates a level of blend of the primary image plane with relation to the background on which the image is being rendered. A common interpretation of the alpha image plane is as a multiplicative processing (normalized to between 0 and 1) applied to the sample values of the primary image plane. The normalized value of the alpha image plane determines the proportion of the blending. The value 0 indicates full transparency and the maximum representable value indicates full opacity.

When NumComponents > 1, the components of the primary image plane corresponding to non-zero indices are defined to be the chroma *components*.

ExtendedHeigh [0] and ExtendedWidth [0] are referred to as the extended image height and width, respectively. Their values are set as HEIGHT_MINUS1 + 1 + TOP_MARGIN + BOTTOM_MARGIN and WIDTH_MINUS1 + 1 + LEFT_MARGIN + RIGHT_MARGIN 23 (Rec. ITU-T, 2016), (ISO/IEC-29199, 2010).

The chroma component array sizes are specified such that ExtendedHeight[i] is equal to ExtendedHeight[1] and ExtendedWidth[i] is equal to ExtendedWidth[1] for all $i > 1$.

The internal color format or INTERNAL_CLR_FMT specifies the size of the chroma component array, for example,

      IF INTERNAL_CLR_FMT is set to YUV420 THEN,
           ExtendedHeigh[1] = ExtendedHeigh[0]/2
           ExtendedWidth[1] = ExtendedWidth[0]/2
      ELSE IF INTERNAL_CLR_FMT is set to YUV422 THEN,
           ExtendedHeigh[1] = ExtendedHeigh[0]
           ExtendedWidth[1] = ExtendedWidth[0]/2
    ELSE
           ExtendedHeigh[1] = ExtendedHeigh[0]
           ExtendedWidth[1] = ExtendedWidth[0]
    ENDIF

Four syntax elements define the image windowing: TOP_MARGIN, BOTTOM_MARGIN, RIGHT_MARGIN, and LEFT_MARGIN and determine the columns and rows of the extended image that are not present in the output image. With respect to a raster scan ordering in the luma array of an image plane, the first TOP_MARGIN rows are not output, nor are the last BOTTOM_MARGIN rows; also, the first LEFT_MARGIN columns are not output, nor are the last RIGHT_MARGIN columns of the luma component. Similarly, the spatially co-located portions of the chroma components are not output, in a manner that retains the ratios between the sizes of the arrays for the chroma components and that of the luma component.

This partitioning of the components into blocks is called a macroblock (MB) partition. The luma component is partitioned horizontally and vertically into an integer number of 16×16 blocks of samples. The chroma components are partitioned into blocks of size 8×8 for 4:2:0 sampling, of size 16×8 for 4:2:2 sampling or of size 16×16 in the default case. Across all the arrays of MB containing the luma and chroma components are spatially co-located.

A frame can be partitioned into rectangular subregions called tiles. A tile is an area of an image that contains rectangular arrays of macroblocks. Tiles enable regions of the image to be decoded without processing the entire image. Tiling adds flexibility to enable different process configurations; memory saving because smaller tile means less horizontal length resulting in less memory; this reduces either the memory accessing or silicon area; also robustness is added because if one error occurs in one tile, it is not propagated to the others (Rec. ITU-T, 2016), (ISO/IEC-29199, 2010).

The encoder selects the number of tiles by setting the HorizontalTileSlices and VerticalTileSlices properties. The minimum tile size is 16×16 pixels. The encoder adjusts the number of tiles to maintain this restriction.

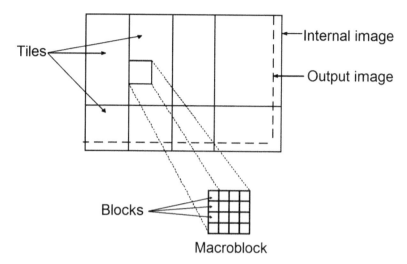

Figure 6.2 Overview of image partitions and internal windowing.

Both the tile width and the tile height correspond to an integer number of macroblocks. The tile partition shall satisfy $1 \leq$ NumTileCols ≤ 4096, and $1 \leq$ NumTileRows ≤ 4096. There is storage and processing overhead associated with each tile; so you should consider the number of tiles that are needed for particular scenarios.

Figure 6.2 is an informative overview of the image plane partitions and implicit windowing, where the extended image plane dimension is indicated by a bold rectangle, the output image plane edges on left and bottom are indicated by dashed lines, 2×4 regular tiling pattern is shown, a macroblock in tile (1,2) is shown, and blocks within the macroblock are shown in an expanded window. Color components are not explicitly shown (Rec. ITU-T, 2016).

The width and height of images are not always integer multiples of 16. When an image width or height is not an integer multiple of 16, the image is extended by some means so that the extended image has dimensions that are multiples of 16. A suggested method of image extension is to perform horizontal replication of the bottom and right edge sample values to the extent necessary to form an input array having a conforming (multiple of 16) extended image width and height.

The JPEG XR standard allows organizing the codestream in two different modes: spatial mode or in frequency mode. In both modes, the image data is organized according to the tile sequence (from left to right and top to bottom) and preceded by the image header information and the index table (containing

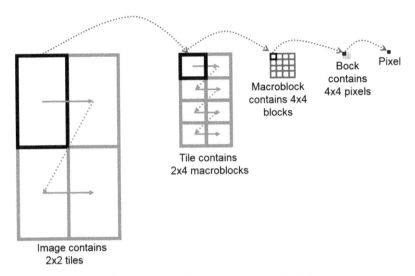

Figure 6.3   Example of image partitioning in JPEG XR.

offset positions of the tiles in the codestream). The modes differ in the way the data is organized inside the tiles.

In spatial mode, the data is organized according to the macroblock sequence from left to right and top to bottom. Figure 6.3 shows an example of how the data is organized in spatial. In the frequency mode, the data is ordered according to their frequency band transmitted according to their importance starting with the low-frequency bands (see Figure 5.9(b)). This allows for a low-quality preview image during transmission similar to progressive JPEG. The next section explains JPEG XR encoder features.

6.3 Encoder

6.3.1 Transform

JPEG XR has half the complexity of JPEG 2000 and preserves the image quality. The transform is a four-channel hierarchical, two-stage, lifting-based, lapped biorthogonal transform (LBT) (Tu, Srinivasan, Sullivan, Regunathan, & Malvar, 2008), (Suzuki & Yoshida, 2017) (See Section 3.6) with a low-complexity structure that makes it exactly invertible in integer arithmetic.

The LBT is implemented as the cascading non-separable 2D transforms of rotation matrices. The core transform is conceptually similar to the widely used DCT and can exploit spatial correlation within the block. The resulting hierarchical two-stage LBT effectively uses long filters for low frequencies and short filters for high frequency detail. The cascaded structures are factorized into lifting structures. In addition, the floating-point lifting coefficients are approximated as dyadic values, as this is of very low cost, thanks to the structures having only adders and shifters without multipliers.

The overlap filtering exploits the correlation across block boundaries and mitigates blocking artifacts. There are three options for overlap filtering: 1) disabled for both stages, 2) enabled for the first stage but disabled for the second stage, or 3) enabled for both stages. Therefore, the transform has a better coding gain as well as reduced ringing and blocking artifacts when compared to traditional block transforms. The overlap filtering is functionally independent of the core transform and can be switched on or off. The option selected by the encoder is signaled to the decoder.

The transform has lossless image representation capability and requires only a small number of integer processing operations for both encoding and decoding. The processing is based on 16 × 16 macroblocks in the transform domain, which may or may not affect overlapping areas in the spatial domain. The design provides encoding and decoding with a minimal memory footprint suitable for embedded implementations.

Each macroblock consists of blocks of 4 x 4 samples, which is the smallest processing unit. In the first transform stage, the lapped transform is applied to individual blocks; the remaining 15 components constitute the high-pass (HP) subband. In the second transform stage, for each macroblock, the transformation procedure groups the 16 DC coefficients of the encompassed 4 × 4 blocks and applies the same transform to this DC-only block. As a result, for each macroblock we have:

- One DC coefficient.
- The low-pass (LP) subband that consists of the 15 non-DC frequency transform coefficients of the DC-only block.
- 240 HP coefficients.

Figure 6.4 shows the frequency hierarchy of a macroblock.

The three subbands (DC, LP, and HP) are quantized and fed to a prediction stage. Prediction residuals then go through coefficient scanning prior to entropy coding.

6.3.2 Steps of transformation

- Optional pre-filter operation (see Figure 6.5), also called photo overlap transform: applied to 4 × 4 areas evenly straddling blocks in two dimensions as follows:
  ○ For images with soft tiles, the filter is applied to all such blocks.

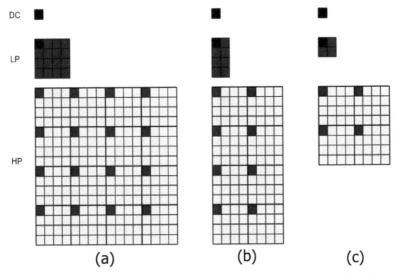

Figure 6.4   Frequency hierarchy of a macroblock. (a) Luma and full resolution chroma. (b) YUV 422. (c) YUV420.

- ○ For images with hard tiles, the filter is applied to the interior of tiles.
- ○ A pre-filter is applied to boundary areas which are 2 × 4 or 4 × 2 in size, and a pre-filter is applied to the four 2 × 2 corners on each color component.
- ○ For images with hard tiles, these filters are additionally applied at tile boundaries.
- A forward core transform (FCT) is applied to all 4 × 4 blocks, yielding one DC coefficient and 15 AC coefficients for each of the 16 blocks in the 16 × 16 region corresponding to the macroblock.
- The resulting 16 DC coefficients within a macroblock are collected into a single 4 × 4 block.
- A second transform stage is applied to this block. This yields 16 new coefficients: one DC component and 15-second stage AC components of this DC block. These coefficients are referred to, respectively, as the DC band and low-pass (LP) band coefficients of the original macroblock.
- The remaining 240 AC coefficients from the first stage transform of the macroblock are referred to as the high-pass (HP) band coefficients.

The chroma components are processed similarly. However, in the case of YUV422 and YUV420, the chroma arrays for a macroblock are 8 × 16 and 8 × 8, respectively. Therefore, the processing performed in the second stage of transformation is adjusted to use a smaller block size. In the YUV420 and YUV422 cases, the chroma component for a macroblock has 60 and 120 HP coefficients, respectively.

The pre-filtering operation is applied across macroblock boundaries and optionally across tile boundaries depending on the status of HARD_TILING_FLAG.

These operations are repeated for all color components. For the special cases of INTERNAL_CLR_FMT equal to YUV422 or YUV420, appropriately modified transforms are applied (for instance, a 2 × 2 transform is used as the block transform of the chroma component DC-LP arrays of YUV420).

The application of overlap filters is determined by the value of OVERLAP_MODE. Lossless coding is possible with all overlap modes, although OVERLAP_MODE equal to 0 is usually best or sufficient for lossless coding. OVERLAP_MODE equal to 1 produces the shortest codestream for a large class of images and quantization levels. OVERLAP_ MODE equal to 2 is recommended for high quantization levels but with slightly higher complexity. OVERLAP_MODE equal to 0 is recommended for lowest complexity encoding/decoding; however, this mode implements a hierarchical block transform and potentially introduces blocking at low bit rates.

The LBT contains the photo core transform (PCT) and the photo overlap transform (POT) (Chengjie, Srinivasan, Sullivan, Regunathan, & Malvar, 2008). The operator of the photo core transform is applied to 4 × 4 pixel blocks within macroblocks to compress the correlation between pixels of each block in the spatial domain, and in the second stage transform, it helps to compress the correlation of all low-pass coefficients of macroblocks in the frequency domain. The PCT is almost the same as the discrete cosine transform DCT since the inverse PCT offers a reliable reconstruction of images transformed by DCT.

Sharp boundaries of blocks appear in lossy JPEG images because block DCT is unable to preserve correlation across block edges. Therefore, the photo overlap transform is designed to hold the correlation between blocks. As shown in Figure 6.5, the operator of POT is translated two pixels backward in both dimensions among blocks.

6.3.3 Quantization

The quantization operation reduces the entropy of the transform coefficient values by rounding them to approximate values. Similar to JPEG (ISO/IEC-10918, 1994), the quantizer uses a scaling factor (SF) or quantization step

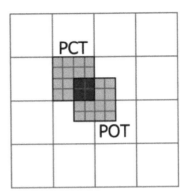

Figure 6.5   Two-stage lapped biorthogonal transform.

to regulate the size of the increment between adjacent selectable decoded coefficient values. The SF is of the form

$$SF = \begin{cases} QP & QP \in [1,16] \\ 2^{\lfloor QP/16 \rfloor - 1}(QP \bmod 16 + 16) & QP \in [17,255] \end{cases}. \tag{6.1}$$

A parameter referred to as the quantization parameter (QP) is used to select the step size. The JPEG XR standard offers quantization parameters from 1 to 255. The scaled coefficient is obtained by dividing the original coefficient by the corresponding SF of the selected quantization parameter and then rounded to an integer.

A small change in QP results in a small change of the quantization step size; for larger values of QP, the same incremental difference results in a larger difference in the quantization step size. The transform coefficient values are integers before quantization. Therefore, for lossless coding, the QP value is 1, and for lossy coding, it is greater than 1.

The DC QP index within a tile is fixed and that across tiles may vary. The HP and LP QP indices within a tile may take on either the same value or one of a multiple value set. The QP index used may change at every macroblock and is specified in the codestream. The following ways of controlling quantization parameters are allowed to deploy various bit allocation techniques to improve the quality of encoded image:

Quantization control on a spatial region basis:
- A global QP value can be selected in the image plane header.
- A QP value can be selected in the tile header to allow different tiles with different QPs.

- Different QP values can be selected for different macroblocks within a tile. Up to 16 different QP values for each LP and HP bands can be selected in the tile header. The QP for DC bands of all the macroblocks in the tile has to be identical.

Quantization control on a frequency band basis:
- All frequency bands have the same QP value.
- Each frequency band has a different QP value.
- DC and LP bands may use the same QP value, whereas HP band can use a different QP value.
- LP and HP bands may use the same QP value, whereas DC band can use a different QP value.

Quantization control on a color plane component basis:
- The QP value, in the uniform mode, is the same for all the colors.
- The QP value, in the mixed mode, for the luma color plane is set to one value, whereas for all other color planes, it is set to a different value.
- The QP value for each color plane can be specified separately in the independent mode.

Quantization control type combinations:
JPEG XR allows for a combination of the above flexibilities.

6.3.4 Prediction of Transform Coefficients and Coded Block Patterns

JPEG XR uses inter-block prediction and coded block patterns to remove the redundancy of the transform coefficients and to gain additional coding efficiency. The four types of inter-block prediction are:

- Prediction of DC coefficient values across macroblocks within tiles.
- Prediction of LP coefficient values across macroblocks within tiles.
- Coded block pattern (CBP) prediction for HP blocks within macroblocks and across macroblocks within tiles.
- Prediction of HP coefficient values within macroblocks.

For example, DC coefficients can be predicted from the top neighbor macroblock, the left neighbor macroblock, and both neighbor macroblocks inside the tile or not be predicted. The prediction is determined by computing the differences of DC coefficients of neighboring macroblocks and no prediction is applied if the difference is large. If DC prediction is applied,

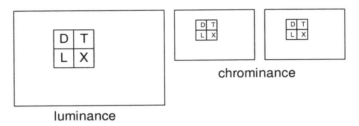

Figure 6.6   DC prediction (Srinivasan *et al.*, 2007).

the top or left LP edges of macroblocks can be predicted by the neighbor macroblocks which are involved in DC prediction as shown in Figure 6.6, where D represents the top-left coefficient (diagonal), T is the top, L is left, and X is the coefficient being predicted.

For LP prediction, the first column can be predicted from its left neighbor macroblock, the first row predicted from its top neighbor macroblock, or no prediction applied. The top or left HP edges of blocks can be predicted by using the neighbor blocks inside macroblocks, which are determined by the LP coefficients. In other words, the LP coefficients either use three coefficients from the top row (1, 2, and 3) of a macroblock or three coefficients from the leftmost column (4, 8, and 12). Figure 6.7 shows the definition and indices of the DC and LP coefficients that are predicted in a 4 × 4 block. The DC coefficient of the block shown in dark gray is at position 0 and the LP coefficients that can be predicted are shown in light gray.

For HP prediction, the first column can be predicted from its left neighbor block, the first row can be predicted from its top neighbor block, or no prediction predicted at all (ISO/IEC-29199, 2010). Figure 6.8 shows a HP left prediction scheme.

0	1	2	3
4	5	6	7
8	9	10	11
12	13	14	15

Figure 6.7   LP coefficients (Srinivasan *et al.*, 2007).

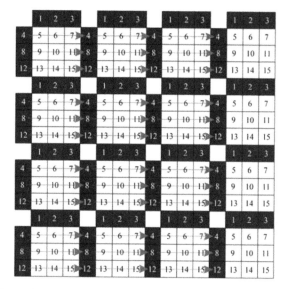

Figure 6.8 HP left prediction scheme (Srinivasan *et al.*, 2007).

Options are identical to LP prediction. However, DC and LP prediction occurs across macroblock boundaries, but HP prediction is only applied from the blocks in the same macroblock. If an individual block does not have a neighbor in the direction specified by prediction mode, no prediction is performed for that block.

For each macroblock, the HP prediction mode is derived from the energy of the LP coefficients of that macroblock. If the energy of the first column of LP coefficients is much smaller than the energy of the first row of LP coefficients, prediction from top is chosen as the HP prediction mode. If the energy of the first row of LP coefficients is much smaller than the corresponding energy of the first column, prediction from left is chosen. Otherwise, no prediction is performed. The energy of the LP coefficients in the chroma channel is also used in deriving HP prediction mode.

6.3.5 Adaptive Scanning Pattern

In order to increase the entropy coding efficiency, it is desirable that these coefficients be scanned in a "descending on the average" order; that means scanning the most probable nonzero coefficients first in an orderly fashion.

In JPEG XR, the scanning patterns are dynamically adapted depending on the local statistics of coded coefficients in the tile. The adaptation process adjusts the scan pattern so that coefficients with higher probability of non-zero values are scanned earlier in the scanning order.

6.3.6 Adaptive Coefficient Normalization (Entropy Coding)

Adaptive coefficient normalization is a type of entropy coding that divides the transform coefficients into a VLC-coded part and a fixed length code (FLC) part (FLEXBITS parameter) in a manner designed to control (i.e., "normalize") bits used to represent the VLC-coded part. The FLC part of DC coefficients and LP coefficients is called FLC refinement and the FLC part of HP coefficients is called flexbits. In other words, VLC is applied to compress the most significant bits (MSB) of the transform coefficients with low entropy and FLC is applied to the least significant bits (LSB) with high entropy and the FLC data are arranged as flexbits band after HP band (ISO/IEC-29199, 2010). One advantage of layering is that flexbits represent a refinement layer which can be used to improve the quality of the decoded image. It may also be omitted or truncated to further reduce the size of the compressed image. This process provides progressive decoding to JPEG XR (Doufaux, Sullivan, & Ebrahimi, 2009).

A few representative VLC tables $\{\mathbf{vT}i : i = 1,\ldots,n\}$ are designed for a wide range of statistics, i.e., tables $\mathbf{vT}1$ and $\mathbf{vT}2$ are more similar to each other than tables $\mathbf{vT}1$ and $\mathbf{vT}3$. After entropy coding a symbol, the adaptation process computes the relative advantage of the two nearest tables for coding this symbol. The most suitable code table is computed using a metric and it is selected based on the history of recently coded symbols. For example, if there exist the nearest VLC tables $\mathbf{vT}k$, $\mathbf{vT}m$, $\mathbf{vT}n$ table $\mathbf{vT}m$ is used for entropy coding of symbol mX. For each value of symbol mX a metric *deltaDisc* is used to estimate the relative cost of coding this symbol using table $\mathbf{vT}k$ or table $\mathbf{vT}n$ instead of table $\mathbf{vT}m$. The cost is stored in tables $\mathbf{vT}m..DeltaTableIndex$ and $\mathbf{vT}m..Delta2TableIndex$. If *deltaDisc* > 0, the symbol is more efficiently coded by the new table instead of the current table. For example, if a given value of the symbol requires 4 bits in the new table and 6 bits in the current table, this value is more efficiently coded in the new table because *deltaDisc* = $6 - 4 = 2$.

The weights obtained from tables $\mathbf{vT}m..DeltaTableIndex$ and $\mathbf{vT}m..Delta2TableIndex$ are added to the discriminants $\mathbf{vT}m..DiscromVal1$ and $\mathbf{vT}m..DiscromVal1$ used for accumulating the statistics regarding code table transition.

A relatively small number of tables is sufficient to provide adaptability to a wide range of image statistics and compression ratios. Sometimes, only two code tables are present and so there is only one possible transition. If only one discriminant is required, the adaptation complexity is further reduced. Therefore, the complexity of the adaptive VLC table algorithm is smaller than the complexity of the actual VLC encoding/decoding algorithm.

6.4 Codestream Structure

JPEG XR standard defines two types of codestreams, the spatial mode and the frequency mode. The spatial mode is a codestream structure mode where the DC, low-pass, high-pass, and flexbits frequency bands for each specific macroblock are grouped together. A single tile packet carries the codestream of each tile in macroblock raster scan order (scanning left to right and top to bottom). The bits associated with each macroblock are located together unlike the frequency mode, where the DC, low-pass, high-pass, and flexbits frequency bands for each tile are grouped separately; the codestream of each tile is carried in multiple tile packets, where each tile packet carries transform coefficients of one frequency band of that tile. The DC tile packet carries information of the DC value of each macroblock, in raster scan order. The LP tile packet carries information of the LP coefficients value of each macroblock. The HP tile packet carries information of the VLC-coded part of the HP coefficients of each macroblock. Finally, the flexbits tile packet carries information regarding the low order bits of the HP coefficients. In both modes, the codestream is composed of a header, followed by a sequence of tile packets.

Figure 6.9 shows the codestream structure. The image header is followed by a sequence of tiles, which are in spatial or frequency mode. The quality

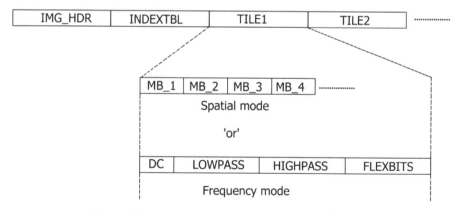

Figure 6.9 Codestream structure (ISO/IEC-29199, 2010).

refinement at the decoding time can be provided at two levels: decoding the codestream without the flexbits, and decoding the entire codestream.

6.5 Profiles

The JPEG XR standard defines a set of capabilities known as profiles, targeting specific classes of applications. Profiles are directly related to the computational complexity. The JPEG XR profiles are listed in Table 6.1 (Rec. ITU-T, 2016).

Sub-Baseline profile: It is a very basic profile for the lowest complexity in both computational and memory requirements.

Baseline profile: This profile defines the constraints for a basic decoder of relatively low complexity in both computational and memory requirements.

Main profile: This profile supports many of the image formats that are supported in the JPEG XR image coding specification and also supports bit depth up to 32 bits with unsigned integer, fixed point signed integer, float, or half formats.

Advanced profile: This profile supports all image formats that are supported in JPEG XR image coding specification. It is intended for decoder implementations that can decode all possible image formats within the JPEG XR design.

Profiles are indicated in the codestream by PROFILE_IDC parameter. Values of PROFILE_IDC that are less than 44 may be associated with the decoder to the Sub-Baseline profile. PROFILE_IDC values less than 55 may be interpreted as expressing conformance of the associated image coded to the baseline profile. PROFILE_IDC values less than 66 may be interpreted as expressing conformance of the associated coded image to the main profile. PROFILE_IDC values less than 111 may be interpreted as expressing conformance of the associated coded image to the Advanced profile.

Table 6.1   JPEG XR profiles (ISO/IEC-29199, 2010), (Rec. ITU-T, 2016).

Profile	Supported features	Superset of
Sub-Baseline	8-bit gray-scale Up to 10-bit per channel RGB	–
Baseline	Up to 16-bit integer and fixed point gray-scale Up to 16-bit integer and fixed point per color RGB	Sub-baseline
Main	Gray-scale, RGB, CMYK, *n*-channel Alpha channel 32-bit integer, fixed point, float32, float16	Sub-baseline and baseline
Advanced	All supported image formats in JPEG XR	Sub-baseline, baseline, and main

6.6 Levels

A level is a set of constraints that indicate a degree of required decoder performance for a profile. Levels constrain the image width and height, the number of tiles horizontally and vertically, the size of tiles horizontally and vertically, and the number of bytes needed to store the decoded image (ISO/IEC-29199, 2010), (Rec. ITU-T, 2016). The level is indicated in the codestream by LEVEL_IDC parameter.

Table 6.2 specifies the levels of JPEG XR. The different values of LEVEL_IDC are specified in the "Level" column. The following restriction shall be observed (ISO/IEC-29199, 2010):

- WIDTH_MINUS1 + LEFT_MARGIN + RIGHT_MARGIN shall be less than MaxImageDimension.
- HEIGHT_MINUS1 + TOP_MARGIN + BOTTOM_MARGIN shall be less than MaxImageDimension.
- NUM_HOR_TILES_MINUS1 shall be less than MaxDimensionInTiles.
- NUM_VER_TILES_MINUS1 shall be less than MaxDimensionInTiles.
- TILE_WIDTH_IN_MB[n] × 16, when present, shall be less than MaxTileDimension for all values of n greater than or equal to 0 and less than NUM_VER_TILES_MINUS1.
- (MBWidth – LeftMBIndexOfTile[NUM_VER_TILES_MINUS1]) x 16 shall be less than MaxTileDimension.
- TILE_HEIGHT_IN_MB[n] * 16, when present, shall be less than MaxTileDimension for all values of n greater than or equal to 0 and less than NUM_HOR_TILES_MINUS1.
- (MBHeight – TopMBIndexOfTile[NUM_HOR_TILES_MINUS1]) * 16 shall be less than MaxTileDimension.
- When LEVEL_IDC is not equal to 255, the value of the function ImageBufferBytes(valNC), as specified in Table B.2, shall be less than

Table 6.2 Levels of JEPG XR (JPEG, Overview of JPEG XR, 2018), (Rec. ITU-T, 2016).

Level	MaxImage dimension	MaxDimension in tiles	MaxTile dimension	MaxBuffer size in bytes
4	2^{10}	2^4	2^{10}	2^{22}
8	2^{11}	2^5	2^{11}	2^{24}
16	2^{12}	2^6	2^{12}	2^{26}
32	2^{13}	2^7	2^{12}	2^{28}
64	2^{14}	2^8	2^{12}	2^{30}
128	2^{16}	2^{10}	2^{12}	2^{32}
255	2^{32}	2^{12}	2^{32}	N/A

MaxBufferSizeInBytes, where the value of valNC is determined as follows:

- ○ If LEVEL_IDC is within a PROFILE_LEVEL_CONTAINER () syntax structure, valNC is set equal to the value in the "NC" column of Table A.6 for the corresponding value of PIXEL_FORMAT.
- ○ Otherwise, if ALPHA_IMAGE_PLANE_FLAG is equal to FALSE, valNC is set equal to the value returned by DetermineNumComponents() for the primary image plane.
- ○ Otherwise, valNC is set equal to the value returned by the function DetermineNumComponents() for the primary image plane plus 1.

In general:

- WIDTH_MINUS1 plus 1 is the output image width.
- HEIGHT_MINUS1 plus 1 is the output image height.
- NUM_VER_TILES_MINUS1 is a 12-bit syntax element that is present when TILING_FLAG is equal to TRUE and specifies the number of tiles in a row minus 1.
- TILE_HEIGHT_IN_MB[n] is a syntax element that specifies the height (in macroblock units) of the nth tile row, where the 0th tile row is the top tile row in the image, and subsequent tile rows are numbered consecutively, top to bottom.
- TILE_WIDTH_IN_MB[n] specifies the width (in macroblock units) of the nth tile column, where the 0th tile column is the leftmost tile column in the image, and subsequent tile columns are numbered consecutively, left to right.
- NUM_HOR_TILES_MINUS1 is a 12-bit syntax element that is present when TILING_FLAG is equal to TRUE and specifies the number of tiles in a column minus 1.
- NUM_VER_TILES_MINUS1 is a 12-bit syntax element that is present when TILING_FLAG is equal to TRUE and specifies the number of tiles in a row minus 1.
- TOP_MARGIN is a 6-bit syntax element that is present when WINDOWING_FLAG is equal to TRUE and specifies the vertical offset of the top boundary of the output image relative to the top edge of the extended image.
- LEFT_MARGIN is a 6-bit syntax element that is present when WINDOWING_FLAG is equal to TRUE and specifies the horizontal offset of the left boundary of the output image relative to the left edge of the extended image.
- BOTTOM_MARGIN is a 6-bit syntax element that is present when WINDOWING_FLAG is equal to TRUE and specifies the vertical offset

of the bottom of the output image relative to the bottom edge of the extended image.

- RIGHT_MARGIN is a 6-bit syntax element that is present when WINDOWING_FLAG is equal to TRUE and specifies the horizontal offset of the right boundary of the output image relative to the right edge of the extended image.
- MBWidth is a variable that holds the value associated with the number of horizontal macroblock partitions.
- MBHeight is a variable that holds the value associated with the number of vertical macroblock partitions.
- LeftMBIndexOfTile[k] is a variable that holds the value associated with the macroblock index of the left macroblock column of the kth tile column.
- TopMBIndexOfTile[i] is a variable that holds the value associated with the macroblock index of the top macroblock row of the ith tile row.

6.7 Decoder

Figure 6.10 shows the block diagram of the decoder with three main stages, the parsing process, and the decoding process for the sample reconstruction.

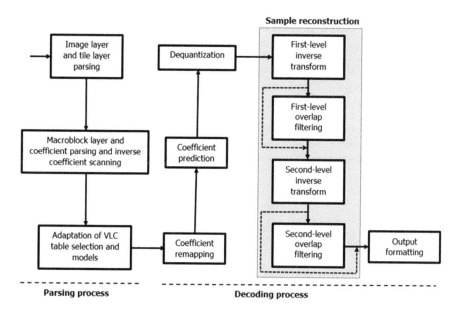

Figure 6.10 Informative decoding process block diagram.

6.7.1 Parsing Process

The codestream consists of the image header, the header of the primary image plane, and, when present, the header of the alpha image plane. The image plane header defines information that is unique to that plane.

A tile index table is used to locate the data that corresponds to a particular tile. A tile packet consists of a tile-packet header, followed by compressed data associated with macroblocks of the tile.

In spatial mode, all the compressed data pertinent to a macroblock is located together in a single tile packet. In frequency mode, each tile packet contains the data associated with a particular transform band; in this mode, a tile packet is classified as a DC tile packet, an LP tile packet, a HP tile packet, or a flexbits tile packet.

If the quantization parameters associated with each band are not specified at the image plane header, they are specified at the tile level.

The macroblock layer is parsed to generate the coefficients of the different frequency bands. These coefficients are inverse transformed to reconstruct the macroblock. Function MB_DC( ) parses the syntax elements related to the DC coefficient, for each component. Function MB_LP( ) parses the syntax elements related to the low-pass coefficients for each component and also performs inverse scanning of the coefficients.

The first step in decoding the HP coefficients involves derivation of the code-block pattern high pass (CBPHP), which determines which 4 × 4 blocks of the macroblock have non-zero coefficients. The CBPHP is parsed as specified by the function MB_CBPHP.

6.7.2 Decoding Process

After the parsing process, the coefficients DC, LP, and HP transform coefficients are remapped. The transform coefficient may be predicted from the coefficients of the neighboring blocks and macroblocks. The dequantization process specifies how the transform coefficients are scaled by the quantizer parameter.

Sample reconstruction:
The inverse transform takes a two-level lapped transform. An inverse core transform (ICT) is applied to each 4 × 4 block corresponding to reconstructed DC and LP coefficients arranged in an array known as the DC-LP array. An overlap filter operation, when indicated, is applied to 4 × 4 areas evenly straddling blocks in the DC-LP array. For images with soft tiles, this filter is applied to all such blocks. For images with hard tiles, this filter is applied only

to the interior of tiles. Furthermore, an overlap filter is applied to boundary 2×4 and 4×2 areas as well as the four 2×2 corner areas. For images with hard tiles, these filters are additionally applied at tile boundaries.

For the parameter INTERNAL_CLR_FMT equal to YUV420 or YUV422, alternate filter operations are applied to the 2×2 interior blocks and 2×1 and 1×2 edge blocks of the chroma components. For these cases, a prediction process is used for the corner samples. The resulting array contains coefficients of the 4×4 blocks corresponding to the first level transform. These coefficients are combined with the reconstructed HP coefficients into a larger array.

The inverse LBT stage includes a two-stage inverse lapped transform. In each stage, an ICT is applied to each 4×4 block. An overlap filter operation, when indicated, is applied to 4×4 areas evenly straddling blocks in the DC-LP array (see Figure 6.10). Two tile types of tile boundary handling are supported: 1) soft tiling in which the transform overlapping stage crosses the tile boundaries and 2) hard tiling in which the transform is applied independently within the individual tiles. For images with soft tiles, the filter is applied to all such blocks. For images with hard tiles, this filter is applied only to the interior of tiles. Furthermore, an overlap filter is applied to boundary 2×4 and 4×2 areas as well as the four 2×2 corner areas. For images with hard tiles, these filters are additionally applied at tile boundaries.

The output formatting accounts for the various transformations required to handle the different color formats and bit depths.

6.8 Motion JPEG XR

Motion JPEG XR is a flexible format, permitting a wide variety of usages, such as editing, display, interchange, and streaming. Motion JPEG XR standard specifies the use of JPEG XR coding for timed sequences of images (ITU-T-T.833).

Motion JPEG XR is expected to be used in a variety of applications, particularly where JPEG XR image coding technology is already available for other reasons, or where the high-quality frame-based approach, with no inter-frame coding, is appropriate.

This Recommendation is voluntary. However, the Recommendation may contain certain mandatory provisions (to ensure, e.g., interoperability or applicability) and compliance with the Recommendation is achieved when all of these mandatory provisions are met.

Motion JPEG XR is expected to be used in a variety of applications, particularly where JPEG XR coding technology is already available for other reasons, or where the high-quality frame-based approach, with no inter-frame coding, is appropriate. These application areas include:

- digital still cameras;
- error-prone environments such as wireless and the Internet;
- video capture;
- high-quality digital video recording for professional broadcasting and motion picture production from film-based to digital systems;
- high-resolution medical and satellite imaging.

Implementations of Motion JPEG XR decoders shall support the decoding of video tracks using JPEG XR coding technology (as specified in Rec. ITU-T T.832|ISO/IEC 29199-2) in files conforming to this International Standard.

6.8.1 Sample Entry and Sample Format

Box types and sample container are "mjxr" and "stbl" for sample table box. The format of a sample when the sample entry name is "mjxr" is a CODED_IMAGE() as defined in (ITU-T-T.832/ISO-IEC-29199-2, 2009). Each image presented to a JPEG XR decoder is logically formed by appending the content of each sample to the content of the JPEG XR Header Box in its associated Visual Sample Entry. Only one CODED_IMAGE( ) is supported per "mjxr" sample entry.

The fields horizresolution and vertresolution in the Visual Sample Entry indicate the highest resolution component of the image (which is typically, but not required to be, the luminance, in an image in which not all components have the same spatial sampling density).

If the coded images contain an alpha plane, a suitable value of "depth," as indicated in the VisualSampleEntry, shall be used.

Color information may be supplied in one or more ColourInformationBox. A ColourInformationBox with an unknown color type may be ignored. The ColourInformationBox is specific to the VideoSampleEntry and should not be confused with the "colr" box defined in other standards such as the JPX file format (see Rec. ITU-T T.801/ISO-IEC-15444-2).

6.8.2 Syntax

```
// Visual Sequences
class MJXRSampleEntry() extends VisualSampleEntry('mjxr')
```

```
{
      JPEGXRInfoBox();
      JPEGXRHeaderBox();
      JPEGXRProfileBox();        // optional
      ColourInformationBox(); // optional
}

class JPEGXRInfoBox() extends FullBox('jxri', 0,
version=0)
{
      UInt8[16] PIXEL_FORMAT;
      UInt8 IMAGE_BAND_PRESENCE;
      UInt8 ALPHA_BAND_PRESENCE;
}
class JPEGXRHeaderBox() extends FullBox('jxrh', 0,
version=0)
{
      IMAGE_HEADER();
      IsCurrPlaneAlphaFlag := FALSE; IMAGE_PLANE_HEADER();
      if (ALPHA_IMAGE_PLANE_FLAG)
      {
         IsCurrPlaneAlphaFlag := TRUE;
         IMAGE_PLANE_HEADER();
      }

}
class JPEGXRProfileBox() extends Box('jxrp')
{
      PROFILE_LEVEL_INFO();
}
class ColourInformationBox extends Box('colr')
{
      unsigned int(32) colour_type;
      if (colour_type == 'nclx') /* on-screen colours */
      {
         unsigned int(16) colour_primaries;
         unsigned int(16) transfer_characteristics;
         unsigned int(16) matrix_coefficients;
         unsigned int(1)  full_range_flag;
         unsigned int(7)  reserved = 0;
         }
```

```
          else if (colour_type == 'rICC')
          {
             ICC_profile; // restricted ICC profile
          }
          else if (colour_type == 'prof')
          {
             ICC_profile; // unrestricted ICC profile
          }
}
```

6.8.3 Semantics

In the Visual Sample Entry:

   Compressorname the value "\016Motion JPEG XR" is suggested but not required (\016 is 14, the length of the string in bytes).

   depth takes one of the following values; other values are reserved, and, if found, the composition behavior is undefined

   0x18 – images are in color with no alpha;

   0x28 – images are in gray-scale with no alpha;

   0x20 – images have alpha (gray or color).

In the JPEG XR Header Box:

IMAGE_HEADER() as defined in clause 8.3 of (ITU-T-T.832/ISO-IEC-29199-2, 2009).

IMAGE_PLANE_HEADER() as defined in clause 8.4 of (ITU-T-T.832/ISO-IEC-29199-2, 2009).

IsCurrPlaneAlphaFlag is not a field in this structure but a local variable used in the decoding of the IMAGE_PLANE_HEADER().

ALPHA_IMAGE_PLANE_FLAG is not a field in this structure but a field in the IMAGE_HEADER() which is tested here PIXEL_FORMAT as defined in clause A.7.18 of (ITU-T-T.832/ISO-IEC-29199-2, 2009).

IMAGE_BAND_PRESENCE as defined in clause A.7.31 of (ITU-T-T.832/ISO-IEC-29199-2, 2009).

ALPHA_BAND_PRESENCE as defined in clause A.7.32 of (ITU-T-T.832/ISO-IEC-29199-2, 2009).

In the JPEG XR Profile Box:
`PROFILE_LEVEL_INFO()` as defined in clause 8.6 of (ITU-T-T.832/ISO-IEC-29199-2, 2009).

In the color information box: For `colour_type` "nclx": these fields are exactly the four bytes defined for `PTM_COLOR_INFO( )` in clause A.7.21 of (ITU-T-T.832/ISO-IEC-29199-2, 2009).

ICC_profile: an ICC profile as defined, e.g., in ISO 15076-1 or ICC.1:2001-04 is supplied.

6.8.4 Profiles and Levels

The conformance to these restricted profiles is indicated in the file type box by the addition of the compatible profiles as brands within the compatibility list. The brand of the Motion JPEG XR Advanced profile shall be in the compatible_brands field of the filetype box (ftyp) in files conforming to this Recommendation (ITU-T-T.833, 2010).

The profile indicators for these profiles are as follows. Since this is a building block specification, these would not normally be used as the major_brand; however, if one of these is the `major_brand`, the `minor_version` must be zero.

Advanced profile "mjxr":
Advanced profile has the following characteristics:
1. At least one video track is present, using at least one MJXRSampleEntryv (see syntax section), (ITU-T-T.833/ISO-IEC-29199-2, 2010).
2. All images conform to the Advanced profile of the JPEG XR image coding specification (ITU-T-T.832/ISO-IEC-29199-2, 2009).

Sub-Baseline profile 'mjxs'
Files conforming Sub-Baseline profile have the following characteristics:
1. All images conform to the Sub-Baseline profile of the JPEG XR image coding specification (ITU-T-T.832/ISO-IEC-29199-2, 2009).
2. Each track shall have exactly one sample description, used by all samples.
3. The file is self-contained; no data references are used, and, therefore, all media data is contained within the single file.
4. The media data in the Media Data Box(es) is placed within the box(es) in temporal order.

If more than one track is present, the media data for the tracks is interleaved, with a granularity no greater than the greater of (a) the duration of a single "sample" (in file format terms) or (b) one second.

6.9 Summary

The JPEG XR specification enables greater effective use of compressed imagery with this broadened diversity of application requirements. JPEG XR supports a wide range of color encoding formats including monochrome, RGB, CMYK and *n*-component encodings using a variety of unsigned integer, fixed point, and floating-point decoded numerical representations with a variety of bit depths. The primary goal is to provide a compressed format specification appropriate for a wide range of applications while keeping the implementation requirements for encoders and decoders simple. A special focus of the design is support for emerging high dynamic range (HDR) imagery applications (JPEG, Overview of JPEG XR, 2018). HDR images require more than the typical 8-bits per sample for the component of a pixel.

Spatial or frequency mode codestreams are available in JPEG XR. In spatial mode, the bits representing different subbands of each tile are packed together in one single data packet that is parsed macroblock-by-macroblock. In frequency mode codestream, the codestream bits for each tile are packed into separate sections – one for each subband. This allows resolution scalability at the decoder. For example, to obtain a 1/256 size image using frequency mode, only the DC data packets need to be decoded; to obtain a 1/16 size image from a frequency mode codestream, only the DC and LP data packets need to be decoded.

To achieve quality scalability, the decoder can decode the portions of codestream corresponding to DC, LP, and HP bands as the first scale of decoding. To enhance the image, the decoder can then proceed to decode the flexbits and add them to create a final version (Rec. ITU-T, 2016). Note that the flexbits can be omitted or truncated to further reduce the size of the compressed image.

The tiling structure of the system allows spatial random access at the decoder side. The granularity of random access is limited by the tile size (large tile size reduces the random access granularity). However, the decoding access complexity varies depending on the shape and location of regions and on the tile size. In other words, more complexity is added when the region of interest (ROI) spans over the border of several tiles. In addition, JPEG XR allows retiling in the compressed domain to get a different access granularity.

Flexbits represents a refinement layer that can be used to improve the quality of the decoded image. Then the portions of codestream corresponding to HP band and flexbits can be eliminated, resulting in a DC and LP-bands-only codestream. This can be done by appropriately changing the BANDS_ PRESENT flag. In addition, the LP band, HP band, and flexbits can be removed. Therefore, the codestream will consist only of DC coefficients.

Spatial mode tiles are laid out in macroblock order, while frequency mode tiles are laid out as a hierarchy of band packets. Both modes can be switched from one to the other, and as entropy coding of coefficients is order-dependent, the codestream must be entropy decoded first in order to entropy code and pack the quantized coefficients in the desired mode.

JPEG XR allows rotation and flip by setting on flag. The flag supports rotation by horizontal and vertical flip as well as rotation by 90° and certain combinations. Besides, it is possible to switch from the interleaved alpha plane mode to a planar mode, by entropy decoding the original codestream, and then entropy encoding image components and the alpha component separately yielding two new codestreams.

7

JPEG XT

The JPEG eXTension or JPEG XT (ISO/IEC 18477) specifies a series of backwards compatible extensions to the legacy JPEG standard (ITU Recommendation T.81|ISO/IEC 10918-1). This coding format allows higher bit depths, HDR (i.e., the use of 32-bit floating-point data for video and image acquisition and manipulation), lossless and near-lossless compression and transparency. JPEG XT is based on a two-layer design, a base layer containing a low-dynamic-range image accessible to legacy implementations, and an extension layer providing the full dynamic range (Richter, 2013).

File formats such as JPEG XR and JPEG 2000 are not backward compatible with the previous JPEG file format. This requires a noticeable investment that induces a difficult transition not always affordable in existing imaging systems. The JPEG XT standard aims to overcome all these drawbacks and to lower the entry barriers to the market and offering new features such WCG and HDR images (Artusi *et al.*, 2016), (Iwahashi & Kiya, 2012), (Iwahashi & Kiya, 2013), (Richter, 2018).

7.1 JPEG XT Parts

JPEG XT currently includes the following nine parts:

Part 1. Core coding system: Specifies the base technology and specifies as such the core JPEG as it is used nowadays, namely as a selection of features from ISO/IEC 10918-1, 10918-5, and 10918-6. Part 1 defines what is commonly understood as JPEG today.

Part 2. Coding of high dynamic range images: It is a backwards compatible extension of JPEG toward high-dynamic range photography using a legacy text-based encoding technology for its metadata. JPEG XT part 2 is a scalable image coding system supporting multiple component images in floating point. It is by itself an extension of the coding tools.

135

Part 3.　　Box file format: Specifies an extensible boxed-based file format all following and future extensions of JPEG will be based on. The format specified in Part 3 is itself compatible to JFIF, ISO/IEC 10918-5, and thus can be read by all existing implementations. Extensions are based on boxes – 64 KB chunks tagged by application marker 11 (APP11), containing enhancement data layers and additional binary metadata describing how to combine them with the base 8-bit layer to form full-precision image.

Part 4.　　Conformance testing: Defines conformance testing of JPEG XT.

Part 5.　　Reference software: Provides the JPEG XT reference software.

Part 6.　　IDR integer coding: Defines extensions of JPEG for backwards compatible coding of integer samples between 9 and 16 bit precision. It uses the file format specified in Part 3.

Part 7.　　HDR floating-point coding: Uses the mechanism of Part 3 to extend JPEG for coding of HDR images, i.e., images consisting of floating-point samples. It is a super-set of both Part 2 and Part 3 and offers additional coding tools addressing needs of low complexity or hardware implementations.

Part 8.　　Lossless and near-lossless coding: Defines lossless coding mechanisms for integer and floating-point samples. It is an extension of Part 6 and Part 7, allowing for scalable lossy to lossless compression.

Part 9.　　Alpha channel coding: Allows the lossy and lossless representation of alpha channels, thus enabling the coding of transparency information and coding of arbitrarily shaped images.

7.2 High Dynamic Range

High dynamic range (HDR) image has larger luminance range than conventional low dynamic range (LDR) image, which is more consistent with human visual system (HVS). HDR images provide a wider range of luminance compared to LDR (Watanabe, Kobayashi, & Kiya, 2018). HDR gives better image quality, but more memory and bandwidth are needed. In order to display an HDR image in an LDR device, a tone-mapping operator (TMO) is used.

Commercial high-end cameras can acquire images with a dynamic range on the order of 10,000:1, quantized into 12–14 bpp; meanwhile, most of the real scene has a dynamic range on the order of 1,000,000:1 or more (Defaux, Le Callet, Mantiuk, & Mrak, 2016). For example, for natural outdoor scenes, the radiances in the image can vary over 4–6 orders of magnitudes, too large to capture by one single typical sensor or represent using a 16-bit integer. According to Defaux, Le Callet, Mantiuk, and Mrak (2016), the three major approaches for capturing HDR video with commercial sensors are:

1. Temporal multiplexing: This technique consists in varying the exposure times of sequences of images from the same scene and then merging them to form the HDR image. However, this methodology requires non-rigid registration of the sequence of images limiting the robustness and the real-time applications.
2. Optical elements to project incidental optical image onto multiple image sensors.
3. Spatial multiplexing: This technique uses a single sensor whose response varies according to the incident light. Optical filter array is used in front of the sensor to vary the spatial response.

Several systems can be used to acquire HDR video; these include optical setups for robust HDR video capture using multiple sensors and spatially varying sensor responses. Conventional digital cameras can be used. HDR images must be taken using sequential images with different exposure. Then, in a post-processing stage, stacked-based algorithms must be used to reconstruct an HDR irradiance image. Currently, two approaches to capture HDR images can be used: HDR sensors or multiple exposure with LDR sensors. Here, there is a need for a standard for coding of HDR images and backward compatible to the popular JPEG compression that allows to be implemented using standard 8-bit JPEG coding hardware or software.

7.3 Encoder

The encoder depends mainly on parameter "q" that controls the base layer coding quality index, which is the same parameter used in JPEG compression tools, and the parameter "Q" that controls the quality of the residual HDR information. Low values of "q" (i.e., $q <= 40$) produce a low quality base layer. A values of "q" and "Q" (i.e., "q" $=99$ and $Q = 99$) yield a near lossless decoded image in profile C.

Typically, HDR images are displayed on legacy monitors using tone-mapping operators (TMO) (i.e., algorithms that map HDR content into the luminance range and color gamut of conventional displays must be used to view the image on LDR monitors), which map HDR wider range of contrasts and colors to the ranges available in the displays. An LDR image is generated by the TMO and then compressed to keep backward compatibility with JPEG. Some common TMOs are "Simple" (a simple linear tone mapping with inverse gamma correction), "Reinhard" (Reinhard, Stark, Shirley, & Ferwerda, 2002), "Drago" (Drago, Myszkowski, Annen, & Chiba, 2003), and "iCAM06" (Kuang, Johnson, & Fairchild, 2007).

The different ways to generate the high dynamic range image is generated from the base image are called profiles. Currently, Part 7 defines four profiles A, B, C, and D (Richter, Artusi, & Ebrahimi, 2016). JPEG XT allows mixing of various elements from different profiles in the codestream, allowing extended DCT precision and lossless encoding in all profiles

7.3.1 Profile A

Figure 7.1 shows the encoding process of Profile A. This profile has the ability to generate better results at low bit rates than the Profile B and has lower computational complexity than the Profile C. The HDR image is described as the pixel-wise product of an inversely gamma-corrected 8-bit tone-mapped image multiplied by a common scale factor. The scale factor is encoded in logarithmic space. Profile A also includes an additive chroma residual that allows correction of color defects. The LDR image is simply tone-mapped HDR image and after color transformation, the components are encoded using baseline JPEG.

The residual layer is generated by subtracting the tone-mapped image from the HDR image normalized by a scale factor which is a function of the luma component of the residual layer. The residual is log-encoded and

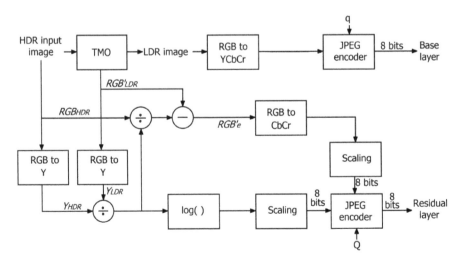

Figure 7.1   JPEG XT profile A encoder.

compressed as an 8-bit gray-scale image. To extend the working gamut, this profile applies a pre-scaling to the tone-mapped image before computing the residual and a post-scaling to the residual.

7.3.2 Profile B

In Profile B, the HDR image is described as the pixel-wise quotient of the inversely gamma-corrected tone-mapped image and a divisor image delivered by the extension layer. The output is additionally scaled by one common factor, the exposure value. This profile is motivated by the following observation: Assume that the base layer consists of a gamma-corrected version of the original HDR image that was clamped to the luminance range available to the legacy JPEG. As long as the pixel values in the HDR image are within this range, the denominator represented by the extension layer can remain one.

If the luminance in the base layer was clamped to the maximum possible value, and the HDR luminance is not representable by means of an 8-bit format, the denominator must fall below one to amplify the output of the JPEG XT decoder to the sample value in the original image. Consequently, for such a simple tone mapping, the extension layer contains all the overcast image regions not carried by the base image, and the base layer contains all regions that are visible under regular exposure. The exposure factor then controls where the cut between regular and extended dynamic range lies. Profile B cannot reach high bit rates and thus cannot reach very high quality in terms of mean squared error. Figure 7.2 depicts the encoder architecture of Profile B.

7.3.3 Profile C

Profile C computes the residual image as a ratio of the HDR image and the inverse tone-mapped image. Unlike the other profiles, the inverse TMO is

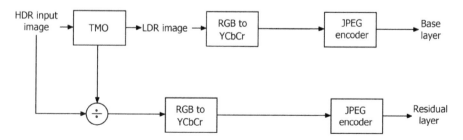

Figure 7.2 JPEG XT profile B encoder.

not a simple inverse gamma, but rather a global approximation of the inverse of the (possibly local) TMO that was used to generate the base-layer image. Similarly, in Profile B, the ratio is implemented as a difference of logarithms. However, instead of using the exact mathematical log operation, Profile C uses a piece-wise linear approximation, defined by re-interpreting the bit-pattern of the half-logarithmic IEEE representation of floating-point numbers as integers, which is exactly invertible. Note that the inverse TMO (ITMO) and log are combined in a single operation and implemented as a lookup table.

Overall, Profile C requires only integer operations, namely, additions, subtractions, and table lookup operations and, hence, allows low complexity hardware implementations. Additionally, Profile C can use a refinement scan in both layers to go from the legacy 8-bit representation to 12-bit (higher precision). The rate-distortion curve of the 8-bit and 12-bit extension modes lie exactly on each other as quantization loss dominates, except that the 12-bit mode allows Profile C to extend this curve toward higher bit rates and higher qualities, allowing scalable lossy to lossless coding. Figure 7.3 depicts the encoder architecture of Profile C.

7.3.4 Profile D

Profile D uses a simple algorithm, which does not generate an enhancement image – the enhancement layer is used to store extended precision of DCT transform coefficients, and non-gamma transfer function is applied to increase

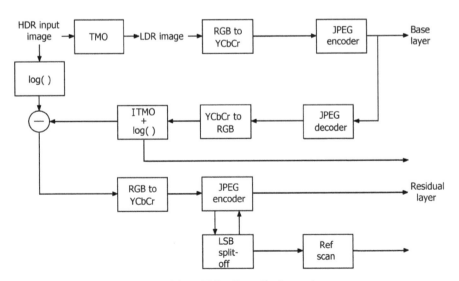

Figure 7.3   JPEG XT profile C encoder.

dynamic range to 12 bits. While Profiles A–C first reconstruct base and extension layers as independent JPEG streams and merge them pixel by pixel in the spatial domain, Profile D extends the resolution of the image precision by extending the precision of the DCT coefficients beyond the resolution allowed by legacy JPEG. The reconstructed image data then undergoes an electro-optical transfer function (EOTF) signaled in the codestream, an idea related to the gamma transformation used in standard dynamic range. Profile D is a very simple entry-level decoder that roughly uses the 12-bit mode of JPEG. Profiles A, B, and C all take into account the nonlinearity of the human visual system. They essentially differ on the strategy used for creating the residual information and on the pre- and post-processing techniques.

7.4 Lossless Coding and Lifting

JPEG XT offers two approaches to enable lossless coding. If 8- to 12-bit precision is sufficient, a lossless integer-to-integer DCT replaces that of legacy JPEG – defined only loosely. Similar to conventional DCT, the lossless DCT is split into a sequence of "elementary rotations" – that is, linear transformations that operate on pairs of coefficients by rotating them in the $\mathbf{R}^2$ space by a step-dependent angle. The lossless DCT replaces each rotation by an equivalent sequence of three "shearing steps," each of which is invertible. This trick, known as "lifting," is also the building block of the wavelet transformation within JPEG 2000. While the original rotation is only invertible up to an implementation-specific precision loss, each sharing depends only on integer operations and is, thus, exactly invertible, resulting in an overall invertible approximation of the rotation.

The second approach also requires fully specifying the DCT process, but, unlike the first, it is based on a lower-complexity fixed-point-based approximation of the DCT. This DCT is, by itself, not completely lossless. Similar to the lossy modes, a second codestream, hidden within application markers, then includes the residual error to compensate for the predictable coding error of the DCT in the base layer. Because the DCT is fully specified, an encoder can predict the errors caused by the DCT in the decoder and compensate for them to allow lossless reconstruction.

A legacy decoder will only recognize the base layer and will reconstruct it by its lossy DCT algorithm. Lossless reconstruction requires a JPEG XT conforming decoder. For that, of course, the error residuals must be decoded without loss, something legacy JPEG cannot ensure. Two possibilities exist: simply bypass the DCT in the extension layer and encode directly in the spatial domain without further transformation or transform the error residuals with the lossless integer-to-integer DCT of the first approach (Richter, Artusi, & Ebrahimi, 2016).

7.5 Coding of Opacity Information

JPEG XT allows the possibility to include a transparency layer by the same means. It includes extension information for higher bit depth. A second codestream, decoding to the opacity information, is hidden within the application markers of the actual foreground image, along with the meta-information that describes how to decode it. Consequently, the image in total consists of two JPEG codestreams: the foreground image and, hidden within it, a one-component JPEG image decoding to the alpha channel.

Although 256 levels of opacity, as granted by a legacy JPEG 8-bit description, is sufficient for many applications, the full feature set of JPEG XT is available to encode the opacity information. Therefore, even floating-point-based opacity samples with a precision of up to 16 bits can be carried in the extended format.

As for all JPEG XT specifications, an image including opacity information is decodable by any JPEG decoder, but only decoders conforming to JPEG XT will pick up the transparency layer (Richter, Artusi, & Ebrahimi, 2016).

Figure 7.4 shows the architecture of the opacity-enabled JPEG XT decoder. The image consists of two JPEG codestreams: the foreground image and, hidden within it, a one-component JPEG image decoding the alpha channel.

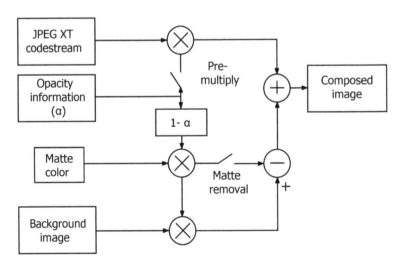

Figure 7.4   Opacity-enabled JPEG XT decoder.

7.6 Privacy and Protection

JPEG privacy and security is a standardization initiative that puts the control of image distribution back into the user's hands. Details on the technology are still being worked out, but the JPEG Committee has already identified a list of requirements that a future JPEG privacy standard must address. JPEG privacy and security will not only allow protection of an image as a whole, it will also allow constrained access to image regions. A typical application is to pixelate only the faces while keeping the rest of the image intact. The resulting privacy-protected image can be distributed without further restrictions, while access to the full image would be limited to a close group of friends (Richter, Artusi, & Ebrahimi, 2016).

The encrypted data must be embedded into the codestream of the scrambled image. JPEG XT already outlines how to hide such additional image data and metadata from legacy decoders, and how to structure the data; JPEG privacy and security will likely build on the same technology used in JPEG XT to store the extension data for HDR photography.

A portion of the image that is encrypted is encoded in the residual layer, and a decryption key is used to make it available to the end user. First, the original image is segmented, manually or automatically, into public and protected regions that require privacy protection or other security mechanisms. Second, the identified image regions must be pixelated or protected. The differential image between the original and scrambled versions contains the pixel-wise difference between the two versions. Third, the data required for decoding the original image (the differential image) should be encrypted by a standard public/private key algorithm. Fourth, the encrypted data must be embedded into the codestream of the scrambled image (Richter, Artusi, & Ebrahimi, 2016).

7.7 Decoder

JPEG XT ensures that existing tools and software will continue to work with the new codestreams, while new features become available that will help move JPEG forward. Backward compatibility, on the other hand, allows new JPEG XT software and hardware to process existing JPEG codestreams. As with any compliant JPEG codestream, it starts with the Start of Image (SOI) marker and ends with the End of Image (EOI) marker. These two markers signal the respective beginning and ending of a single encoded image in a codestream. The first segment contains an encoded base layer for an LDR image reconstruction (i.e., 8-bit per component). Although not strictly required by the JPEG XT specification, this base layer typically contains a tone-mapped version of the original image at a specific quality level, as would be expected from decoding any regular JPEG photograph. The set of 16 APP marker segments was originally defined to allow embedding of third-party data in JPEG codestreams. Unknown APP markers are ignored by the decoder. While legacy decoders will only see the base layer image data, JPEG XT decoders will interpret the extra data from the APP11 marker segments.

The second segment contains JPEG XT extension data embedded in one or more APP11 marker segments. These segments can carry any type of byte-oriented content. The extension data is structured using generic data containers called boxes that carry the type of the box and the body to inform the actual payload. Type is a four-byte identifier to inform the decoder its purpose and the structure of the respective body content. Boxes are embedded in the codestream format by encapsulating them into one or several JPEG XT marker segments. A single box may extend over multiple JPEG XT marker segments and merge multiple marker segments before decoding the box content.

JPEG XT marker segments that belong to the same logical box and, thus, require merging prior to interpretation, will have identical box instance number fields (En), but different packet sequence number fields (Z). The packet sequence number is a 32-bit field that specifies the order in which payload data shall be merged. Figure 7.5 shows the complete JPEG XT marker segment structure (APP11 in JPEG decoders). The Le field represents the marker segment length (16 bits), while the LBox field represents the total logical box size (32 bits). The common identifier (CI) field is always set to 0x4A50 ("JP"). The actual box payload data depends on the box type and is defined in the various JPEG XT specifications.

Figure 7.6 shows the functional block diagram of the standard decoder from parts 3–8. The JPEG XT image consists of a legacy codestream using the 8-bit Huffman coding mode of the legacy coding system and the residual codestream inserted in applications markers to extend the target precession.

APP11	Le	CI	En	Z	LBox	TBox	XLBox	DBox

APP11	marker 0xFFEB	LBox	Box length
Le	Size of marker segment	TBox	Box type
CI	Common identifier for specific APP11 marker	XLBox	Box length extension
En	Box instance number	DBox	Payload data
Z	Paquet sequence number		

Figure 7.5   Organization of the APP11 marker segment for boxes.

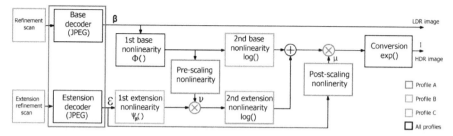

Figure 7.6 The simplified JPEG XT standard decoder architecture. The gray block is the legacy decoding system (JPEG).

The extension decoder in the bottom row works likewise and uses either an 8-bit or a 12-bit mode of the JPEG standard (Artusi *et al.*, 2016).

In Profile A, the HDR image I is represented as a product of a luminance scale μ and the base image B after inverse gamma correction through $\Phi(\bullet)$ (first base nonlinearity). μ is a scalar function of the luma component extension image in the post-scaling nonlinearly block (Artusi *et al.*, 2015). The reconstruction algorithm for profile A can be expressed as

$$I(x,y) = \mu(\varepsilon_0(x,y)) \cdot \left[C\,\Phi(B(x,y)) + \upsilon\big(SC\Phi(B(x,y))\big) \cdot R\varepsilon^{\perp}(x,y) \right] \quad (7.1)$$

where C and R are 3×3 matrices implementing color transformation. The matrix C transforms from ITU-R BT.601 to the target colorspace in the base image. If the target color space is, for example, ITU-R BT.2020, this matrix can be computed as

$$C = \begin{bmatrix} 1.544 & -0.320 & -0.228 \\ -0.567 & 1.375 & 0.013 \\ 0.023 & -0.055 & 1.214 \end{bmatrix} \begin{bmatrix} 0.640 & 0.290 & 0.150 \\ 0.330 & 0.600 & 0.060 \\ 0.030 & 0.110 & 0.790 \end{bmatrix} \quad (7.2)$$

where the first matrix is the conversion from XYZ to linear BT.2020, and the second matrix is the conversion from linearized BT.601 to XYZ.

R is an inverse decorrelation transformation from YCbCr to RGB in the extension image. Typically, it is identical to the conversion from YCbCr to RGB as defined by BT.601

$$R = \begin{bmatrix} 1.000 & 0.000 & 1.402 \\ 1.000 & -0.344 & -0.714 \\ 1.000 & 1.172 & 0.000 \end{bmatrix} \quad (7.3)$$

S is a row-vector transforming color into luminance, and v is a scalar function of this luminance value. Typically, $v(x) = x + \varepsilon$ where ε is a "noise floor" that avoids an instability in the encoder for very dark image regions. Inverting the decoder Equation (7.1) as necessary for encoding includes a division by $v(SB)$. Dark image areas result in denominators to come close to zero.

The reconstruction algorithm for each of the ith RGB color channels of profile B is

$$I(x,y)_i = \sigma \cdot \exp\left(\log\left(\left[C\Phi(B(x,y)) \right]_i \right) - \log\left(\psi\left(\left[R\varepsilon(x,y) \right]_i \right) + \varepsilon \right) \right)$$

$$I(x,y)_i = \sigma \cdot \frac{\left[C\Phi(B(x,y)) \right]_i}{\psi\left(\left[R\varepsilon(x,y) \right]_i \right) + \varepsilon} \qquad (i = 0,1,2). \qquad (7.4)$$

The HDR image I is, in general, represented as the RGB component-wise quotient B/ε where ε is the unity in areas that are captured in the LDR base image. In this profile, $\Phi(\bullet)$ is the inverse gamma correction and $\psi(\bullet)$ is the content-dependent nonlinearity applied in the extension. The exponential and logarithmic functions are realized by the second base, second extension nonlinearities, and output conversion blocks. The additional scale σ can be understood as an exposure parameter that scales the luminance of the output image to optimize the split between base and extension images. Note that this profile follows a different strategy by splitting the image along the luminance axis into "overexposed" areas and LDR areas (Artusi *et al.*, 2015).

In profile C, $\Phi(\bullet)$ approximates an inverse gamma transformation and carries out a global inverse tone-mapping procedure that approximates the (possibly local) TMO used to create the LDR image. The extension is encoded in the logarithmic domain directly, avoiding an additional transformation. Finally, log and exp are substituted by piece-wise linear approximations that are implicitly defined by re-interpreting the bit-pattern of the half-logarithmic IEEE representation of floating-point numbers as integers. The reconstruction algorithm is

$$I(x,y) = \psi \cdot \exp\left(\check{\Phi}(CB(x,y)) + R\varepsilon(x,y) - 2^{15}[1,1,1]^T \right) \qquad (7.5)$$

where $\check{\Phi}(x) = \psi \log(\Phi(x))$, where Φ is the global inverse tone mapping approximation. 2^{15} is an offset shift to make the extension image symmetric around.

Profile C allows lossless coding as its decoding algorithm only requires invertible integer operations; the only change in the lossless mode is that the DCT in the reconstruction of ε is bypassed, and the implementation of the base layer DCT is fully specified by the standard. The refinement scans increase the bit-precision in the DCT domain (Artusi *et al.*, 2015).

7.8 Summary

JPEG XR is a coded image format standardized by ITU-T and ISO/IEC. Its design is based on a Microsoft-developed technology known as HD Photo and it addresses the needs of a broad range of consumer electronics applications, particularly including digital photography (Richter, Bruylants, Schelkens, & Ebrahimi, 2018).

JPEG XR uses a hierarchical, two-stage lapped biorthogonal transform (LBT). Each LBT consists of a concatenation of two operators: the photo core transform (PCT), similar to the DCT, and the photo overlap transform (POT).

As in JPEG 2000, tiling is the primary mechanism provided in the JPEG XR design to enable fast access to spatial regions. JPEG XR and JPEG 2000 additionally support access to different levels of an image resolution hierarchy and can provide scalable fidelity enhancement data (Hiroyuki & Hitoshi, 2017).

In terms of objective quality, JPEG 2000 performs slightly better than JPEG XR, in accordance with the subjective tests. As expected, the wavelet-based codec receives higher scores in comparisons to the other one.

JPEG XT is backward compatible to the legacy JPEG and encodes images of higher precision and higher dynamic range in lossy or lossless modes. Embedding mechanism used in JPEG XT is possible, thanks to a legacy JPEG structure called "application marker." This means that any JPEG decoder will be able to decode a JPEG XT file. JPEG XT offers other features such as the inclusion of opacity information (alpha channels), privacy and security protection of image content or image regions, animated JPEG, and efficient coding of omnidirectional (360°) images (Artusi *et al.*, 2015).

The first backward compatible of JPEG XT to legacy JPEG is to enable coding of images with bit depths higher than 8 bits per component. Unlike an image coded in the little known 12-bit mode of JPEG, a legacy JPEG decoder only implementing the popular 8-bit mode would still be able to understand a JPEG XT codestream, though the output would then only have a precision of 8 bits per sample.

JPEG XT does this by first encoding an 8-bit version of the high-precision input, also called base layer, and hiding a second codestream known as enhancement layer, within this legacy codestream that enlarges its precision to a fuller range; up to 16 bits per component or 48 bits in total as currently specified. Additional metadata, also embedded in the legacy codestream, tell a JPEG XT decoder how to combine the base layer and the enhancement layer to form one single image of a higher precision.

8

JPEG 2000

The JPEG 2000 (ISO/IEC-15444, 2004) offers an extremely high level of scalability and accessibility because it uses compression techniques based on the wavelet transform. The codestream is embedded, that is, content can be coded once at any quality, up to lossless, but accessed and decoded at a potentially very large number of qualities and resolutions and/or by ROI, with no significant penalty in coding efficiency.

The standard supports up to 16,384 components, with dimensions running into the thousands of tera pixels, and precisions as high as 38 bits/sample, with or without tiling, and with a variety of interchangeable data progressions and random access capabilities. The JPEG 2000 architecture lends itself to a wide range of uses from portable digital cameras through to advanced pre-press, medical imaging, geospatial, and other key application domains. The requirements from different application areas may be summarized as follows.

- Improved low bit-rate performance: It should give acceptable quality below 0.25 bpp (networked image delivery and remote sensing applications).
- Lossless and lossy compression: Lossless compression included in progressive decoding. It supports embedded bit stream for progressive lossy to lossless build-up.
- Progressive transmission: The standard allows progressive transmission for images to be reconstructed with increasing pixel accuracy and resolution.
- Region of interest coding: JPEG 2000 allocates more bits to the ROIs as compared to the non-ROI areas.
- Random codestream access: The user-defined ROIs in an image are randomly accessible.
- Error resilience: The standard should provide error resiliency, especially for wireless communication channels.

149

- Open architecture: It should support open architecture to optimize the system for different image types and applications. The decoder should have only the core tool set and a parser that understands the core stream. If necessary, the unknown tools can be requested by the decoder.
- Content-based description: Finding the desired image from a large archive of images is a challenging task. This has applications in medical images, forensic, digital libraries, etc. These issues are being addressed by MPEG-7.
- Image security: This is another major requirement to protect the intellectual property rights of the images. Digital images can be protected using watermarking, labeling, stamping, encryption, etc.
- Continuous-tone and bi-level image compression: Previously, we had different standards for bi-level images (JBIG standards) and continuous-time images (JPEG standard). A requirement was felt that a single standard should address both these domains, including the color images and encompass applications involving compound images with overlaying texts.
- Sequential one-pass decoding: The decoding should be done in a single sequential pass and support interleaved and non-interleaved modes.
- Side channel support: The standard should support alpha planes and transparency planes.

8.1 JPEG 2000 Parts

The JPEG 2000 image coding system (JPEG, Overview of JPEG 2000, 2018), (ISO/IEC 15444) consists of following parts:

Part 1.   Core coding system: Defines the core of JPEG 2000: The syntax of a JPEG 2000 codestream and the necessary steps involved in decoding JPEG 2000 images, with informative guidance for encoders. Part 1 also defines a basic file format called JP2, which allows metadata such as color space information and IP rights to be provided with the JPEG 2000 codestream. The standardization committee developed the Part 1 of JPEG-2000 standard to make it royalty free so as to promote its wide usage.

Part 2.   Extensions: Defines codestream and file format extensions including: multi-component transformations; more flexible wavelet transform kernels and decomposition structures; alternate quantization schemes; and nonlinear point transforms. The Part 2 JPX file format extends the Part 1 JP2 file format to allow: more comprehensive color space descriptions and HDR sample representations; multiple codestreams; composition, cropping, geometric transforms; rich animations; descriptive metadata; and a rich metadata set for photographic imagery.

Part 3. Motion JPEG 2000 (MJ2 or MJP2): Defines a file format for motion sequences of JPEG 2000 images, where each image is coded as an independent JPEG 2000 codestream.

Part 4. Conformance: Specifies test procedures for both encoding and decoding processes defined in Part 1, including the definition of a set of decoder compliance classes. The Part 4 test files include both bare codestreams and JP2 files.

Part 5. Reference software: Consists of two source code packages that implement Part 1. The implementations were developed alongside Part 1, and were used to test it. One is written in C and the other in Java. They are both available under open-source licenses.

Part 6. Compound image file format: Defines the JPM file format for multi-page document imaging, which uses the mixed raster content (MRC) model of ISO/IEC 16485. JPM is an extension of the JP2 file format defined in Part 1 (LSoft-Technologies, 2019). Although it is a member of the JPEG 2000 family, it supports the use of many other coding or compression technologies, including JBIG2 and JPEG.

Part 7. Abandoned: Guideline of minimum support function of Part 1.

Part 8. JPEG 2000 secured (JPSEC): Standardizes tools to ensure the security of transaction, protection of contents (IPR), and protection of technologies (IP), and to allow applications to generate, consume, and exchange JPEG 2000 secured bit streams.

Part 9. JPIP: Defines tools for supporting incremental and selective access to imagery and metadata in a networked environment. A primary focus for Part 9 is efficient and responsive interactive remote browsing of JPEG 2000 content conforming to any of the other parts of the standard.

Part 10. JP3D: Is the volumetric extension of JPEG 2000 Part 1. It explicitly defines the notion of an extra spatial dimension (the Z-dimension), extending key JPEG 2000 concepts such as tiles, precincts, and code-blocks to all three dimensions so as to provide resolution- and region-of-interest accessibility properties in 3D. Part 10 also adds support for wavelet decomposition structures that extend hierarchically in all three dimensions.

Part 11. JPWL: Defines tools and methods to achieve the efficient transmission of JPEG 2000 imagery over an error-prone wireless network. More specifically, Part 11 extends the elements in the core coding system described in Part 1 with mechanisms for error protection and correction. These extensions are backward compatible: Decoders which implement Part 1 are able to skip the extensions defined in Part 11.

Part 12.   Withdrawn in 2017: ISO base media file format

Part 13.   Entry-level encoder: Defines an entry-level encoder implementation of Part 1.

Part 14.   JPXML: Specifies an XML representation of the JPEG 2000 file format and marker segments, along with methods for accessing the internal data of a JPEG 2000 image.

Part 15.   HTJ2K (Under-development): Specifies an alternate block coding algorithm that can be used in place of the existing block coding algorithm specified in Part 1. The alternate block-coding algorithm is intended to offer a tenfold increase in throughput at the expense of slightly reduced coding efficiency, while allowing mathematically lossless transcoding to/from codestreams that use the Part 1 block-coding algorithm and preserving Part 1 codestream syntax and features.

8.2 Architecture of JPEG-2000

Figure 8.1 shows the block-diagram of JPEG-2000 (a) encoder and (b) decoder. The core transform used is the discrete wavelet transform (DWT), rather than the DCT used in JPEG. Before applying the DWT, the source image is divided into components (colors) and each component is divided into tiles which are compressed independently, as though they are independent images and all the samples in the tile are DC level shifted. The DWT coefficients in different subbands are quantized and then composed into an embedded bit stream following the Embedded Block Coding with Optimized Truncation (EBCOT) algorithm. The embedded bit stream is composed of quality layers so as to offer both resolution and SNR scalability. The entropy coding is based on the MQ coder, which has error correction capability.

8.3 Reference Grid

The codec describes the geometry of the various components in terms of a rectangular grid of size $Xsiz \times Ysiz$ called the reference grid as shown in Figure 8.2. Each dimension of the reference grid cannot exceed the size $2^{32} - 1$ units. This is an upper bound on the size of an image that can be

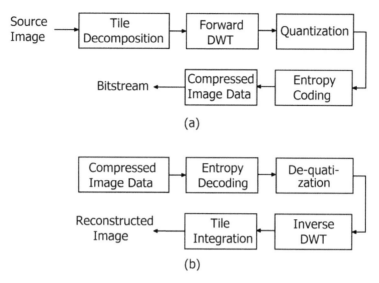

(a)

(b)

Figure 8.1 JPEG 2000 codec. (a) Encoder. (b) Decoder.

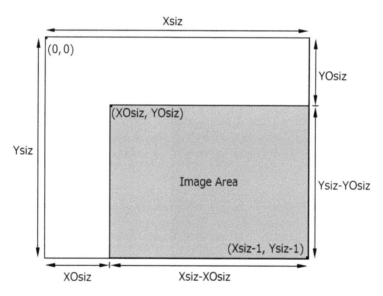

Figure 8.2 Reference grid.

handled by the codec. All of the components are mapped onto the image area of the reference grid. Consequently, additional information is required in order to establish this mapping because components need not be sampled at the full resolution of the reference grid (Adams, 2013).

The image area or picture data to be represented is the rectangle described by the left-top coordinate (XOsiz, YOsiz) and the right-bottom coordinate (Xsiz-1, Ysiz-1). For each component, the horizontal and vertical sampling periods must be indicated in units of the reference grid, denoted as XRsiz and YRsiz, respectively which specify a rectangular sampling grid consisting of all points whose horizontal and vertical positions are integer multiples of XRsiz and YRsiz, respectively. Thus, in terms of its own coordinate system, a component will be of the size

$$\left(\left\lceil\frac{Xsiz}{XRsiz}\right\rceil - \left\lceil\frac{XOsiz}{XRsiz}\right\rceil\right) \times \left(\left\lceil\frac{Ysiz}{YRsiz}\right\rceil - \left\lceil\frac{YOsiz}{YRsiz}\right\rceil\right). \tag{8.1}$$

Its top-left simple will be the point

$$\left(\left\lceil\frac{XOsiz}{XRsiz}\right\rceil, \left\lceil\frac{YOsiz}{YRsiz}\right\rceil\right). \tag{8.2}$$

For a given image, many combinations of the Xsiz, Ysiz, XOsiz, and YOsiz parameters can be chosen to obtain an image area with the same size.

8.4 Tiling Grid and Tiles

Before applying the DWT, the image and its components are divided into non-overlapping blocks. Each of these blocks is called a tile. For an image with multiple components, each tile also consists of these components. For a gray-scale image, the tile has a single component.

Each tile may be coded independently, as if each tile is an independent image. Therefore, all operations, such as component mixing, DWT, quantization, and entropy coding are carried out independently for each tile.

Tiling reduces memory requirements for DWT and its processing and is amenable to parallelization. A tile size of 256×256 or 512×512 is found to be a typical choice for VLSI implementations.

In addition, tiles can be independently accessed and used for decoding specific parts of the image, rather than the complete one. However, in terms of PSNR, tiling degrades the performance, as compared to no tiling and smaller tile sizes lead to tiling artifacts.

Tiling is performed with respect to the reference grid by overlying a rectangular grid called tiling grid. The tiling grid is aligned with the point (XTOsiz, YTOsiz) with horizontal and vertical grid spacing of XTsiz and YTsiz, respectively, and overlaid on the reference grid as shown in Figure 8.3.

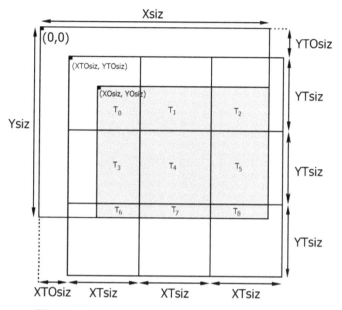

Figure 8.3   Reference grid, tiling grid, and image area.

The tile sizes can be arbitrary up to the size of the original image (that is, a single tile). All the tiles have the same nominal dimension XTsiz × YTsiz, except the tiles at the image boundary if the dimension of the image is not an integer multiple of the dimension of the tiles (Adams, 2013).

The partitioning of the components is obtained by mapping the position of each tile, from the reference grid, to the coordinate systems of the individual components. For example, suppose that the upper-left and lower-right corners of a particular tile has coordinates (tx_0, ty_0) and $(tx_1 - 1, ty_1 - 1)$, respectively. Then, in the coordinate space of a particular component, the tile would have an upper-left corner and lower-right corner with coordinates (tcx_0, tcy_0) and $(tcx_1 - 1, tcy_1 - 1)$ where

$$\left(tcx_0, tcy_0\right) = \left(\lceil tx_0/XRsiz \rceil, \lceil ty_0/YRsiz \rceil\right) \qquad (8.3)$$

$$\left(tcx_1, tcy_1\right) = \left(\lceil tx_1/XRsiz \rceil, \lceil ty_1/YRsiz \rceil\right). \qquad (8.4)$$

Equations (8.3) and (8.4) correspond to the tile-component coordinate system in Figure 8.4. The portion of a component that corresponds to a single tile is referred to as a tile-component.

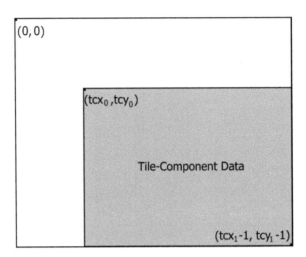

Figure 8.4   Tile-component coordinate system.

Although the tiling grid is regular with respect to the reference grid, it is important to note that the grid may not necessarily be regular with respect to the coordinate systems of the components (Adams, 2013).

8.5 Image Pre-Processing

The image to be encoded may comprise from 1 component, as in the gray-scale images, to 2^{14} components. Each component is signed or unsigned integer-valued samples with a precision from 1 to 38 bits. The various components of an image need not be sampled at the same resolution. Thus, the components can have different sizes (see Section 2.8.1).

For unsigned input sample data values in the range $\left[0, 2^{B} - 1\right]$, each sample is centered by subtracting the same quantity 2^{B-1} to each sample value, where B is the precision of a component in bits/sample, signaled in the SIZ (image and tile size) marker segment in the compressed bit stream. For images whose samples are represented by signed integers in the range $\left[-2^{B-1}, 2^{B-1} - 1\right]$, as in the case of some type of medical images, the dynamic range is already centered about zero, and no DC level shifting is required.

8.6 Forward Intercomponent Transform

After applying the pre-processing stage, the forward intercomponent transform can be applied to the tile-component data. This step reduces the correlation among components, leading to improved coding efficiency.

For color compression, the images are converted to YCbCr space and then center each of the Y, Cb, and Cr channels. The JPEG 2000 Part 1 standard supports two different transformations: RCT for both lossless and lossy compression because the errors are introduced by the transformation and/or quantization steps only, not by the RCT, and irreversible color transform ICT applied in only lossy compression. ICT introduces errors due to forward and inverse color transformation by using non-integer coefficients as the weighting parameters in the transformation matrix (see Section 2.8).

8.7 Forward Intracomponent Transform

In this stage, transforms that operate on individual components can be applied. The particular type of operator employed for this purpose is the wavelet transform.

Fourier analysis expands an arbitrary signal in terms of infinite number of sinusoidal functions of its harmonics and has been well studied by the signal processing community for decades. Fourier representation of signals is known to be very effective in analysis of time-invariant (stationary) periodic signals. However, Fourier analysis does not allow both time and frequency representation of signal simultaneously.

A wavelet is a small wave whose energy is concentrated in a short time; their properties allow both time and frequency representation of signals simultaneously because wavelets are small oscillations that are highly localized in time and still possesses the wave-like (periodic) characteristics.

8.7.1 The Continuous Wavelet Transform

The continuous wavelet transform (CWT) basis functions are scaled (dilated) and shifted versions of the time-localized mother wavelet $\psi(t)$ are as follows:

$$\psi_{a,b}(t) = \frac{1}{\sqrt{|a|}} \psi\left(\frac{t-b}{a}\right) \tag{8.5}$$

where variables a and b are integers that represent the scaling and shifting parameters. It is obvious that the mother wavelet is obtained when $a = 1$ and $b = 0$. The Morlet–Grossmann definition of the continuous wavelet transform (Grossmann & Morlet, 1984) for a 1D signal $f(t) = L^2(\mathbb{R})$ is

$$W(a,b) = \frac{1}{\sqrt{|a|}} \int_{-\infty}^{\infty} f(t) \cdot \psi_{a,b}^*(t)\, dt \tag{8.6}$$

where $\psi^*(\bullet)$ denotes complex conjugation. The parameter a causes contraction of $\psi(t)$ in the time axis when $a < 1$ and expansion or stretching when $a > 1$. The inverse wavelet transform is

$$f(t) = C^{-1} \int_{a=-\infty}^{\infty} \int_{b=-\infty}^{\infty} \frac{1}{|a|^2} \, W(a,b) \cdot \psi_{a,b}^*(t) \, da \, dt \tag{8.7}$$

where

$$C = \int_{-\infty}^{\infty} \frac{|\Psi(\Omega)|^2}{|\Omega|} \, d\Omega \tag{8.8}$$

with $\psi(\Omega)$ being the Fourier transform of the mother wavelet. The CWT maps a one-dimensional function $f(t)$ to a function $W(a,b)$ of two continuous real variables a and b. The admissible constant states that wavelet does not have a zero frequency component; therefore, it has zero mean.

$$C = \int_{0}^{\infty} \frac{|\Psi(\Omega)|^2}{|\Omega|} \, d\Omega < \infty. \tag{8.9}$$

The discrete wavelet transform (DWT) can be defined in terms of discrete values of the dilation and translation parameters a and b instead of being continuous. The most popular approach of discretizing a and b is

$$a = a_0^m, \qquad\qquad b = nb_0 a_0^m \tag{8.10}$$

where m and n are integers. Replacing Equation (8.10) into Equation (8.5), we obtain

$$\psi_{m,n}(t) = a_0^{-m/2} \, \psi\left(a_0^{-m} t - nb_0\right). \tag{8.11}$$

There are many choices to select a_0 and b_0. The most common choices are $a_0 = 2$ and $b_0 = 1$. This way of sampling is popularly known as *dyadic sampling* and the corresponding decomposition of the signals is called the *dyadic decomposition*. This corresponds to sampling (discretization) of a and b in such a way that the consecutive discrete values of a and b as well as the sampling intervals differ by a factor of two. Hence, the following discrete wavelets constitute a family of orthonormal basis functions

$$\psi_{m,n}(t) = 2^{-m/2} \, \psi\left(2^{-m} t - n\right). \tag{8.12}$$

The discrete wavelet coefficients of a function $f(t)$ can be derived accordingly as

$$c_{m,n}(f) = 2^{-m/2} \int f(t)\psi\left(2^{-m}t - n\right) dt. \tag{8.13}$$

To recover the signal from the discrete wavelet coefficients

$$f(t) = \sum_{m=-\infty}^{\infty}\sum_{n=-\infty}^{\infty} c_{m,n}(f)\,\psi_{m,n}(t). \tag{8.14}$$

In general, the wavelet coefficients for a function $f(t)$ are given by

$$c_{m,n}(f) = a_0^{-m/2} \int f(t)\psi\left(a_0^{-m}t - nb_0\right) dt. \tag{8.15}$$

Equation (8.15) is called the wavelet series similar to the Fourier series. The input function $f(t)$ is still a continuous function, whereas the transform coefficients are discrete. This is often called the discrete-time wavelet transform (DTWT). When the input function $f(t)$ as well as the wavelet parameters a and b are represented in discrete form, the transformation is commonly referred to as the discrete wavelet transform (DWT) of signal $f(t)$ and is the basis of the JPEG 2000 image compression standard.

8.7.2 Multiresolution Analysis

A multiresolution representation (Haddad & Akansu, 2012) of an image gives us a complete idea about the extent of the details existing at different locations from which we can choose our requirements of desired details. Multiresolution representation facilitates efficient compression by exploiting the redundancies across the resolutions. Wavelet transforms is one of the popular, but not the only, approaches for multiresolution image analysis.

Besides the mother wavelet, for multiresolution analysis, a scaling function $\phi(t)$ is defined. For example, any function $f(t)$ can be analyzed as a linear combination of real-valued expansion functions $\phi_m(t)$ as

$$f(t) = \sum_{n=-\infty}^{\infty} \alpha_n\, \phi_{m,n}(t) \tag{8.16}$$

where n is an integer index of summation (finite or infinite), the α_ns are the real valued expansion coefficients, and $\{\phi_m(t)\}$ forms an expansion set. The set of expansion functions is represented by the integer translations and binary scaling of the real, square-integrable function $\phi(t) \in L^2(\mathbb{R})$ of the following equation:

$$\phi_{m,n}(t) = 2^{-m/2} \; \phi\left(2^{-m} t - n\right). \tag{8.17}$$

For fixed m, the set of scaling functions $\phi_{m,n}(t)$ are also orthonormal. The set of functions $\left\{\phi_{m,n}(t)\right\}$ can be made to cover entire square-integrable real space $L^2(\mathbb{R})$. Note that if we make $m = m_0$, the set of function $\left\{\phi_{m_0,n}(t)\right\}$ obtained by integer translations of n can only cover one subspace of the entire $L^2(\mathbb{R})$ which is the subspace spanned V_0 defined as the functional subspace of $\left\{\phi_{m_0,n}(t)\right\}$ at a given scale m_0. The width of the set of functions $\left\{\phi_{m_0,n}(t)\right\}$ is twice of that of the set of functions $\left\{\phi_{m_0+1,n}(t)\right\}$. Hence, the functional subspace spanned by $\left\{\phi_{m_0+1,n}(t)\right\}$ contains the subspace spanned by $\left\{\phi_{m_0,n}(t)\right\}$. In other words, the subspace spanned by the scaling functions at lower scales is contained within the subspace spanned by those at higher scales and is given by the following nested relationship and shown in Figure 8.5(a)

$$\{0\} \subset \cdots \subset V_{-1} \subset V_0 \subset V_1 \subset V_2 \subset \cdots \subset L^2(\mathbb{R}). \tag{8.18}$$

Note that

$$\phi(t) = \sqrt{2} \sum_n h_\phi(n)\phi(2t - n) \tag{8.19}$$

where $h_\phi(n)$ are the scaling function coefficients. The scaling functions and the wavelet functions differ by their spanning subspaces. W_0 is the subspace spanned by $\left\{\psi_{m_0,n}(t)\right\}$; then,

$$V_m = \mathrm{Span}\left\{\phi_{m,n}(t)\right\} \tag{8.20}$$

$$W_m = \mathrm{Span}\left\{\psi_{m,n}(t)\right\}. \tag{8.21}$$

A set of wavelet functions span the difference subspace between two adjacent scaling function subspaces V_m and V_{m+1}. Therefore, the scaling functions and the wavelet functions differ by their spanning subspaces. We can write the multiresolution analysis in terms of wavelet functions as

$$L^2(\mathbb{R}) = V_0 \oplus W_0 \oplus W_1 \oplus \cdots. \tag{8.22}$$

Also,

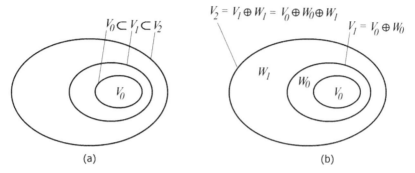

Figure 8.5 Nested function spaces spanned by (a) the scaling function and (b) relationship between the scaling and wavelet function spaces, where $\oplus$ indicates union of subspaces.

$$L^2(\mathbb{R}) = \cdots \oplus W_{-2} \oplus W_{-1} \oplus W_0 \oplus W_1 \oplus W_2 \oplus \cdots \qquad (8.23)$$

where the symbol $\oplus$ denotes orthogonal sum. Therefore, if we have a function that belongs to the space V_{m-1} (i.e., the function can be exactly represented by the scaling function at resolution $m-1$), we can decompose it into a sum of functions starting with lower-resolution approximation followed by a sequence of functions generated by dilations of the wavelets that represent the loss of information in terms of details as shown in Figure 8.5(b).

Since wavelet spaces reside within the spaces spanned by the next higher scaling functions, any wavelet function can be expressed as a weighted sum of shifted double-resolution scaling functions as follows:

$$\psi(t) = \sqrt{2} \sum_n h_\psi(n)\phi(2t - n) \qquad (8.24)$$

where $h_\psi(n)$ are the wavelet function coefficients. Suppose that a function $f(t)$ to be analyzed belongs to the subspace V_0 but not V_1. The scaling functions V_0 make a crude approximation of the function and the wavelet functions W_0 provide the details. In this sense, the scaling functions analyze $f(t)$ into its low-pass filtered form and the wavelet functions analyze $f(t)$ into its high-pass filtered form.

The major difference between the CWT and discrete wavelet transforms is how the scale parameter is discretized. The CWT discretizes the scale more finely than the discrete wavelet transform.

The scaling functions provide low-frequency analysis of the signals to obtain approximations and the wavelet functions provide high-frequency analysis of the signals to extract the details.

8.7.3 The Discrete Wavelet Transform (DWT)

Let $f(x)$; $x \in 0,1,\dots,M-1$ denote a discrete function. The coefficients of series expansions are as follows.

Approximation coefficients:

$$W_{\phi}(m_0,n) = \frac{1}{\sqrt{M}} \sum_{x} f(x)\, \phi_{m_0,n}(x). \qquad (8.25)$$

Detail coefficients:

$$W_{\psi}(m,n) = \frac{1}{\sqrt{M}} \sum_{x} f(x)\, \psi_{m,n}(x) \quad m \geq m_0. \qquad (8.26)$$

The corresponding inverse discrete wavelet transform (IDWT) to express the discrete signal in terms of the wavelet coefficients can be written as

$$f(x) = \frac{1}{\sqrt{M}} \sum_{n} W_{\phi}(m_0,n)\phi_{m_0,n}(x) + \frac{1}{\sqrt{M}} \sum_{m=m_0}^{\infty} \sum_{n} W_{\psi}(m,n)\psi_{m,n}(x). \qquad (8.27)$$

8.7.4 1D Filter Bank

DWT and subband decomposition have a strong relationship and it is possible to compute DWT through subband decomposition. We first substitute the definition of wavelet functions $\psi_{m,n}(x)$, given in Equation (8.12), into Equation (8.26), to obtain

$$W_{\psi}(m,n) = \frac{1}{\sqrt{M}} \sum_{x} f(x)\, 2^{-m/2}\, \psi\left(2^{-m}x - n\right). \qquad (8.28)$$

We rewrite the expression for wavelet function in terms of weighted sum of shifted, double-resolution scaling function, given in Equation (8.24)

$$\psi(x) = \sqrt{2} \sum_{n} h_{\psi}(n)\phi(2x - n).$$

If we scale x by 2^m, translate it by k, and let $r = 2k + n$,

$$\psi(2^{-m}x - k) = \sqrt{2} \sum_{n} h_{\psi}(n)\phi\left(2\left(2^m x - k\right) - n\right)$$

$$\psi(2^{-m}x - k) = \sqrt{2} \sum_{r} h_{\psi}(r - 2k)\phi\left(2^{m+1}x - (2k + n)\right)$$

$$\psi(2^{-m}x-k)=\sqrt{2}\sum_{r}h_{\psi}(r-2k)\phi\left(2^{m+1}x-r\right). \tag{8.29}$$

Substituting the above expression in Equation (8.28) with $n = k$,

$$W_{\psi}(m,n)=\frac{1}{\sqrt{M}}\sum_{x}f(x)\,2^{-m/2}\left[\sqrt{2}\sum_{r}h_{\psi}(r-2n)\phi\left(2^{m+1}x-r\right)\right].$$

Interchanging the order of summation, we obtain

$$W_{\psi}(m,n)=\sum_{r}h_{\psi}(r-2n)\left[\frac{1}{\sqrt{M}}\sum_{x}f(x)\,2^{-(m-1)/2}\,\phi\left(2^{m+1}x-r\right)\right]. \tag{8.30}$$

The term within the square bracket is identified as the DWT equation for the approximation coefficient with m replaced by $(m - 1)/2$; therefore, we can write

$$W_{\psi}(m,n)=\sum_{r}h_{\psi}(r-2n)W_{\phi}(m+1,n). \tag{8.31}$$

Similarly, if we use the definition of scaling functions and apply the weighted summation expressions, it is possible to obtain in a similar manner as above

$$W_{\phi}(m,n)=\sum_{r}h_{\phi}(r-2n)W_{\phi}(m+1,n). \tag{8.32}$$

Note that $W_{\psi}(m,n)$ and $W_{\phi}(m,n)$ are obtained by convolving the approximation coefficients $W_{\phi}(m+1,n)$ with filters $h_{\psi}(-n)$ and $h_{\phi}(-n)$ respectively. Then, the outputs of the filters are down-sampled by a factor of 2 as shown in Figure 8.6.

The DWT can be computed by iterating the approximation coefficients as shown in Figure 8.7.

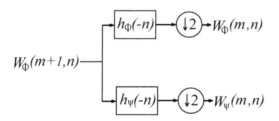

Figure 8.6   One-stage, 1D analysis of DWT coefficients.

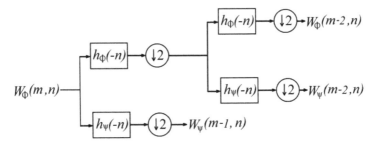

Figure 8.7 Two-stage, 1D analysis of DWT coefficients.

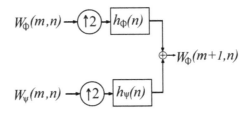

Figure 8.8 One-stage, 1D synthesis of DWT coefficients.

An equally efficient approach exists to implement IDWT through the synthesis of the subband signals, as shown in Figure 8.8.

8.7.5 2D Filter Bank

The one-dimensional DWT can be easily extended to two-dimensional signals. We require to extract the information along rows and along columns (horizontal and vertical directions). In this case, we require to multiply separable scaling and wavelet functions in both directions as $\phi(n_1, n_2) = \phi(n_1)\phi(n_2)$, $\psi^H(n_1, n_2) = \psi(n_1)\phi(n_2)$, $\psi^V(n_1, n_2) = \phi(n_1)\psi(n_2)$, and $\psi^D(n_1, n_2) = \psi(n_1)\psi(n_2)$, where the superscripts H, V, and D stand for horizontal, vertical, and diagonal details. The one-stage, 2D analysis structure is shown in Figure 8.9.

Each subband, corresponding to the filter output, contains one-fourth of the total number of samples as shown in Figures 8.10 and 8.11.

The 2D analysis filter bank can be applied to the approximation coefficients (LL subband) further to have a second level of DWT decomposition as shown in Figure 8.11.

DWT and subband decomposition have a strong relationship and it is possible to compute DWT through subband decomposition.

The LL subband at the highest level of decomposition has significant information content and all other subbands have less significant content. This

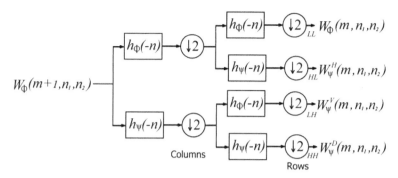

Figure 8.9 One-stage, 2D analysis of DWT coefficients.

Figure 8.10 One level of 2D-DWT decomposition. Upper-left subbands are the approximation coefficients, upper-right subbands are the horizontal detail coefficients, lower-left subbands are the vertical detail coefficients, and lower-right subbands are the diagonal detail coefficients.

Figure 8.11 Two levels of 2D-DWT decomposition.

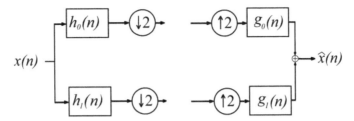

Figure 8.12   Two-band analysis filter bank.

is expected since all natural images are rich in low-frequency information content, as compared to the high-frequency content and the results demonstrate excellent energy compaction properties of DWT. This property can be very effectively utilized by the wavelet-based image coding standards which, after subband decomposition, quantization, and entropy coding strategies, are applied to produce the bit stream. At the decoder side, the subbands are restored and IDWT is applied to yield an approximation of the original image.

8.7.6 Two-Band Analysis and Synthesis Filter Bank

Note that in the analysis bank, $h_\phi(-n) = h_0(n)$ and $h_\psi(-n) = h_1(n)$ are a low-pas and a high-pass filter, respectively. Because of the down-sampling factors, the cutoff frequency of the filters is $\pi/2$ as shown in Figure 8.12.

z-transform of the analysis bank:
The output of the analysis filters (low-pass and high-pass filters) in the z domain is

$$H_0(z)X(z)$$

$$H_1(z)X(z).$$

Hence, the z-transform at the output of the down-samplers is

$$\frac{1}{2}\left[H_0(z^{1/2})X(z^{1/2}) + H_0(-z^{1/2})X(-z^{1/2})\right]$$

$$\frac{1}{2}\left[H_1(z^{1/2})X(z^{1/2}) + H_1(-z^{1/2})X(-z^{1/2})\right].$$

z-transform of the synthesis bank:

The output of the down-samplers is the input of the up-samplers. Then, the output of the up-samplers is

$$\frac{1}{2}\big[H_0(z)X(z)+H_0(-z)X(-z)\big]$$

$$\frac{1}{2}\big[H_1(z)X(z)+H_1(-z)X(-z)\big].$$

These outputs are multiplied by the z-transform of the synthesis filters as

$$\frac{1}{2}\big[H_0(z)X(z)+H_0(-z)X(-z)\big]G_0(z)$$

$$\frac{1}{2}\big[H_1(z)X(z)+H_1(-z)X(-z)\big]G_1(z).$$

The output of the filter bank is given by

$$\hat{X}(z)=\frac{1}{2}\big[H_0(z)X(z)G_0(z)+H_0(-z)X(-z)G_0(z)\big]$$

$$+\frac{1}{2}\big[H_1(z)X(z)G_1(z)+H_1(-z)X(-z)G_1(z)\big].$$

Re-arranging the terms so that we obtain the terms aliased

$$\hat{X}(z)=\frac{1}{2}\big[H_0(z)G_0(z)+H_1(z)G_1(z)\big]X(z)$$

$$+\frac{1}{2}\big[H_0(-z)G_0(z)+H_1(-z)G_1(z)\big]X(-z).$$

For error-free reconstruction, we must ensure that

$$H_0(z)G_0(z)+H_1(z)G_1(z)=2$$
$$H_0(-z)G_0(z)+H_1(-z)G_1(z)=0.$$

In matrix form

$$\big[G_0(z)\quad G_1(z)\big]\begin{bmatrix}H_0(z) & H_0(-z)\\ H_1(z) & H_1(-z)\end{bmatrix}=\big[2\quad 0\big] \tag{8.33}$$

$$[G_0(z) \quad G_1(z)]\mathbf{H}_m(z) = [2 \quad 0].$$ (8.34)

Assuming $\mathbf{H}_m(z)$ is non-singular, then

$$\begin{bmatrix} G_0(z) \\ G_1(z) \end{bmatrix} = \left(\mathbf{H}_m^T(z)\right)^{-1} \begin{bmatrix} 2 \\ 0 \end{bmatrix} = \frac{2}{\det\left(\mathbf{H}_m(z)\right)} \begin{bmatrix} H_1(-z) \\ -H_0(-z) \end{bmatrix}.$$ (8.35)

For finite impulse response (FIR) filters, the determinant of the matrix $\mathbf{H}_m(z)$ is a pure delay given by $c\,z^{-(2k+1)}$, with c being a constant.

Let $c = 2$ and neglect the delay

$$\begin{bmatrix} G_0(z) \\ G_1(z) \end{bmatrix} = \begin{bmatrix} H_1(-z) \\ -H_0(-z) \end{bmatrix}.$$ (8.36)

Therefore,

$$g_0(n) = (-1)^n h_1(n)$$

$$g_1(n) = (-1)^{n+1} h_0(n).$$

Hence, for error-free reconstruction, the FIR synthesis filters are cross-modulated copies of analysis filters, with one of the signs reversed. Note that all the filters can be derived from one filter resulting in the same filter length. When filters on analysis side and synthesis side are same (i.e., same wavelet function with same scaling parameter), then these filter banks are called "Orthogonal Filter Banks."

Since the bandwidth of the signal at each of the analysis filter outputs is only one-half of the original signal, half of the samples at the analysis filter outputs are redundant, and, hence, every one out of two consecutive samples at the filter output can be dropped by using a down-sampler. This is a shortcoming of this scheme because the filters must operate at high rate even though half of the samples are discarded.

8.7.7 Biorthogonality

Filter banks with different analysis and synthesis wavelets and scaling function are called "Biorthogonal Filter Banks." We define the product filter as $P(z) = G_0(z)H_0(z)$. From Equation (8.35), we have

$$P(z) = \frac{2}{\det\left(\mathbf{H}_m(z)\right)} H_0(z)H_1(-z).$$ (8.37)

Note that $\det\left(\mathbf{H}_m(z)\right) = -\det\left(\mathbf{H}_m(-z)\right)$. Therefore, from Equation (8.35), we can also write

$$P(-z) = G_1(z)H_1(z) = -\frac{2}{\det\left(\mathbf{H}_m(z)\right)}H_0(-z)H_1(z) \qquad (8.38)$$

$$P(-z) = G_0(-z)H_0(-z). \qquad (8.39)$$

For error-free reconstruction, we can rewrite

$$G_0(z)H_0(z) + G_0(-z)H_0(-z) = 2. \qquad (8.40)$$

Taking the inverse transform in both sides of Equation (8.40), we obtain

$$\sum_k g_0(k)h_0(n-k) + (-1)^n \sum_k g_0(k)h_0(n-k) = 2\,\delta(n). \qquad (8.41)$$

When n is odd, the terms get canceled and only the even-indexed terms add up. Hence, we can write

$$\sum_k g_0(k)h_0(2n-k) = \delta(n). \qquad (8.42)$$

Also, using Equation (8.39), we can write

$$G_1(z)H_1(z) + G_1(-z)H_1(-z) = 2. \qquad (8.43)$$

Therefore,

$$\sum_k g_1(k)h_1(2n-k) = \delta(n). \qquad (8.44)$$

From alias cancelation, we can use

$$H_0(-z)G_0(z) + H_1(-z)G_1(z) = 0$$

to derive the conditions

$$\sum_k g_0(k)h_1(2n-k) = 0 \qquad (8.45)$$

$$\sum_k g_1(k)h_0(2n-k) = 0. \qquad (8.46)$$

Finally, we can express the biorthogonality condition as

$$\langle g_i(k), h_r(2n-k) \rangle = \delta(i-r)\,\delta(n), \quad i,r = 0,1$$

where $\langle\,,\,\rangle$ is the inner product. Biorthogonality allows us to have filters with different lengths.

8.7.8 Irreversible and Reversible Wavelet Filters

An irreversible scheme (real-to-real mapping) does not allow perfect reconstruction of the input image. This scheme is used for JPEG 2000 lossy encoding and implemented using the 9/7 Daubechies filters shown in Tables 8.1 and 8.2.

In a reversible scheme (integer-to-integer mapping), exact reconstruction is possible at the decoder side and it is included for lossless JPEG 2000 (Taubman & Marcellin, JPEG 2000: Image compression fundamentals, practice and standards, 2001). This is implemented using a 5/3 Le-Gall filters, whose analysis and synthesis filter coefficients are shown in Tables 8.3 and 8.4.

Table 8.1   Daubechies 9/7 analysis filter coefficients.

n	Low-pass filter	High-pass filter
0	0.6029490182363579	1.115087052456994
±1	0.2668641184428723	−0.5912717631142470
±2	−0.07822326652898785	−0.05754352622849957
±3	−0.01686411844287495	0.09127176311424948
±4	0.02674875741080976	

Table 8.2   Daubechies 9/7 synthesis filter coefficients.

n	Low-pass filter	High-pass filter
0	1.115087052456994	0.6029490182363579
±1	0.5912717631142470	−0.2668641184428723
±2	−0.05754352622849957	−0.07822326652898785
±3	−0.09127176311424948	0.01686411844287495
±4		0.02674875741080976

Table 8.3   Le-Gall 5/3 analysis filter coefficients.

n	Low-pass filter	High-pass filter
0	6/8	1
±1	2/8	−1/2
±2	−1/8	

Table 8.4 Le-Gall 5/3 synthesis filter coefficients.

n	Low-pass filter	High-pass filter
0	1	6/8
±1	1/2	−2/8
±2		−1/8

8.7.9 Convolution- and Lifting-Based Filters

Two filtering approaches, namely convolution- and lifting-based filters, are used in JPEG 2000 standard (Taubman & Marcellin, JPEG 2000: Image compression fundamentals, practice and standards, 2001). The signal is extended at the boundaries using symmetric extension to avoid artifacts. The convolution-based filtering performs a dot product between the two filter masks (for low-pass and high-pass filters) and the extended 1D signal.

In lifting-based filtering, the input samples are split into odd and even groups. The odd samples are updated with a weighted sum of even sample values and the even samples are updated with a weighted sum of odd sample values. Figure 8.13 shows the general block diagram of the lifting realization of a 1D two-channel perfect reconstruction (PR) uniformly maximally decimated (UMD) filter bank (FB) (a) analysis and (b) synthesis sides, respectively.

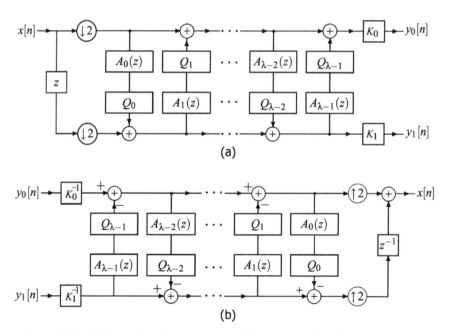

Figure 8.13 Lifting realization of one-stage, 1D, two-channel perfect reconstruction uniformly maximally decimated filter bank (UMDFB). (a) Analysis side. (b) Synthesis side.

The wavelet transform is calculated by recursively applying the 2D UMDFB to the low-pass subband signal obtained at each level in the decomposition. Suppose that an $(R - 1)$-level wavelet transform is to be employed. To compute the forward transform, we apply the analysis side of the 2D UMDFB to the tile-component data in an iterative manner, resulting in a number of subband signals being produced.

Each application of the analysis side of the 2D UMDFB yields four subbands as shown in Figures 8.10 and 8.11. Therefore, the wavelet decomposition can be associated with data at R different resolutions. JPEG 2000 standard supports up to 32-level 2D DWT from the one-dimensional, two-channel subband transforms.

Typical values are 4–8 with 5 levels sufficient for near optimal compression performance. A single level applies the subband transform separately to the columns and then to the rows of the image. If the quantization and coding operations specified by the JPEG 2000 standard will be performed directly on the image sample values, $x[n]$, the DWT decompositions is skipped by setting the number of levels to zero. This is useful when compressing bi-level images or pelletized color images.

In Figure 8.12, $\{A_i(z)\}_{i=0}^{\lambda-1}$, $\{Q_i(x)\}_{i=0}^{\lambda-1}$, and $\{K_i\}_{i=0}^{1}$ denote filter transfer functions, quantization operators, and (scalar) gains, respectively. For integer-to-integer mappings, the $\{Q_i(x)\}_{i=0}^{\lambda-1}$ are selected to yield integer values and $\{K_i\}_{i=0}^{1}$ are integers. For real-to-real mappings, the $\{Q_i(x)\}_{i=0}^{\lambda-1}$ are selected to be the identity and the scalars $\{K_i\}_{i=0}^{1}$ are real numbers.

The 9/7 transform is non-reversible and real-to-real with parameters:

$$\lambda = 4,$$

$$A_0(z) = \alpha_0(z+1),\ A_1(z) = \alpha_1(1+z^{-1}),\ A_2(z) = \alpha_2(z+1),\ A_3(z) = \alpha_3(1+z^{-1})$$

$$\alpha_0 = -1.586134342059924,\quad \alpha_1 = -0.052980118572961,$$

$$\alpha_2 = 0.882911075530934,\quad \alpha_3 = 0.443506852043971$$

$$Q_i(x) = x \quad \text{for } i = 0,1,2,3$$

$$K_0 = 1/1.230174104914001,\quad K_1 = -1/K_0$$

The 5/3 transform is reversible and has parameters

$$\lambda = 2,$$

$$A_0(z) = -\frac{1}{2}(z+1),\ A_1(z) = \frac{1}{4}(1+z^{-1}),$$

$$Q_0(x) = -\lfloor -x \rfloor,\ Q_1(x) = \lfloor x+1/2 \rfloor,$$

$$K_0 = 1,\ K_1 = 1.$$

The number of resolution levels is a parameter of each transform. A typical value for this parameter is 6 (for a sufficiently large image). The encoder may transform all, none, or a subset of the components. At the decoder side, the inversion process is not guaranteed to be exact unless reversible transforms are employed. The lifting is used when reversible transformation is desired. In addition, it requires less memory and fewer arithmetic operations than in the convolution case.

8.8 Quantization

JPEG 2000 offers several different quantization options. Only uniform scalar (fixed-size) dead-zone quantization is included in Part I of the standard. Part II of the standard generalizes this quantization method to allow more flexible dead-zone selection. Furthermore, trellis coded quantization (TCQ) is offered in Part II as a value-added technology (ISO/IEC-15444-1, 2004).

A very desirable feature of compression systems is the ability to successively refine the reconstructed data, as the bit stream is decoded. In this situation, the decoder reconstructs an approximation of the reconstructed data after decoding a portion of the compressed bit stream. As more of the compressed bit stream is decoded, the reconstruction quality can be improved, until the full quality reconstruction is achieved upon decoding the entire bit stream.

Since this property is one of the key focuses of JPEG 2000, the quantizers used in the standard have been designed to enable embedded quantization. In the irreversible case, quantization compresses a range of values to a single quantum value (Marcellin *et al.*, 2002). After the tile-component data has been transformed (by intercomponent and/or intracomponent transforms), the resulting coefficients are quantized.

Part 1 of JPEG 2000 standard (Marcellin *et al.*, 2002) requires that transform coefficients be quantized using scalar quantization with a deadzone (midtread) twice as wide as the other quantization intervals. Quatizers with different step size parameters (Δ_b) are employed for the coefficients $y_b(u,v)$ of each subband b. The definition of the quantization process can be expressed as

$$q_b(u,v) = \text{sign}(y_b(u,v)) \left\lfloor \frac{|y_b(u,v)|}{\Delta_b} \right\rfloor \tag{8.47}$$

where

$$\Delta_b = 2^{R_b - \varepsilon_b} \left(1 + \frac{\mu_b}{2^{11}} \right) \tag{8.48}$$

where $q_b(u,v)$ is the quantizer index value and R_b is the number of bits representing the nominal dynamic range of the subband. Δ_b is represented with a total of 2 bytes, an 11-bit mantissa μ_b, and 5-bit exponent ε_b. When 5/3 filter bank is used, then $\mu_b = 0$ and $\varepsilon_b = R_b$; therefore, $\Delta_b = 1$, and $q_b(u,v)$ will have M_b bits fully decoded. Note that $M_b = G + \varepsilon_b - 1$ and G is the number of guard bits signaled to the decoder (typically one or two).

Two modes of signaling Δ_b to the decoder are used. The expounded quantization where the encoder sends explicitly the (ε_b, μ_b) value for every subband and the derived quantization mode where a single value (ε_0, μ_0) is sent for the LL subband and the values of (ε_b, μ_b) for the remaining subbands are derived by scaling the Δ_0 value by some power of 2 as $(\varepsilon_b, \mu_b) = (\varepsilon_0 - N_L + n_b, \mu_0)$, where N_L is the total number of decomposition levels and n_b is the decomposition level of the subband b. At the decoder side, the transform coefficients are reconstructed using the following equation:

$$Rq_b(u,v) = \begin{cases} \left(q_b(u,v) + \gamma\right)\Delta_b, & \text{if } q_b(u,v) > 0, \\ \left(q_b(u,v) - \gamma\right)\Delta_b, & \text{if } q_b(u,v) < 0, \\ 0, & \text{otherwise,} \end{cases} \qquad (8.49)$$

with $0 < \gamma < 1$ being an arbitrary parameter chosen by the decoder. A popular choice is $\gamma = 0.375$. If M_b bits for a quantizer index are fully decoded, then the step size is equal to Δ_b; otherwise, the step size in Equation (8.49) is $\Delta_b = 2^{M_b - N_b}$, where N_b is the number of decoded bits. In the reversible case, the same happens (with $\Delta_b = 1$), except when the index is fully decoded for lossless reconstruction.

8.9 Partitioning of Subbands

Each subband is divided into non-overlapping rectangles. Three spatially consistent rectangles – one from each subband at each resolution level comprise a packet partition location or precinct, as illustrated in Figure 8.14. Each precinct is further divided into code-blocks, which form the input to the arithmetic coder. The following definitions are important.

• Precinct: Each subband is divided into rectangular blocks called precincts. Precinct sizes can vary from resolution to resolution but restricted to be a power of 2. Precincts represent a coarser partition of the wavelet coefficients than code-blocks, and they provide an efficient way to access spatial regions of an image. The precinct and code-block partitions are both anchored at (0, 0).

Subband

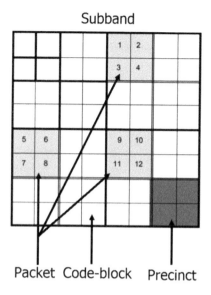

Packet Code-block   Precinct

Figure 8.14 Precincts, packets, and code-blocks.

- Packets: Three spatially consistent rectangles comprise a packet. Within a packet, code-blocks are coded in raster order.
- Code-block: Each precinct is further divided into non-overlapping rectangles called code-blocks. Each code-block forms the input to the entropy encoder and is encoded independently. The coefficients in a code-block are separated into bitplanes. The individual bitplanes are coded in 1–3 coding passes.

8.10 Entropy Coding

The quantized indices, corresponding to the coefficients of each subband, are entropy coded to produce the bit stream, which is possible by using bitplane encoding, encoded 1 bit at a time, starting with the MSB and proceeding to the LSB. In JPEG 2000, the subbands are encoded independently to avoid the problems that face models that exploit the correlation between subbands to improve coding efficiency (Shapiro, 1993), (Said & Pearlman, 1996).

JPEG 2000 uses the embedded block coding with optimized truncation (EBCOT) algorithm (Taubman & Marcellin, 2001), where each subband is divided into non-overlapping rectangles or code-blocks with a minimum height or width size of 4, a maximum number of coefficients not exceeding

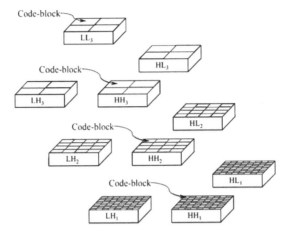

Figure 8.15   Division of subbands into code-blocks of same dimensions in every subband.

4095 per code-block, and the size an integer power of two. This is illustrated in Figure 8.15.

Each code-block B_i is coded independently to produce an elementary embedded bit stream c_i represented by a prefix of length L_i at the corresponding rate. However, there exist a collection of truncation points $L_i^{(z)}$ at which distortion $D_i^{(z)}$ stays the closest to the convex part of the rate-distortion graph, i.e., for a code-block B_i, with a finite number $Z_i + 1$ of truncation points of lengths $L_i^{(z)}$ we have $0 = L_i^{(0)} \le L_i^{(1)} \le \cdots \le L_i^{(z_i)}$.

The overall reconstructed image distortion can be represented as the sum of distortions from each code-block with $D_i^{(z)}$ being the distortion contributed by block B_i if its elementary embedded bit stream is truncated to length $D_i^{(z)}$. We are free to select any set of truncation points such that

$$\sum_i L_i^{(z_i)} \le L_{\max}. \tag{8.50}$$

Of course, the best choice is that which minimized the overall distortion

$$D = \sum_i D_i^{(z_i)}. \tag{8.51}$$

The selection of truncation points may be deferred until after all of the code-blocks have been compressed. This strategy is known as post-compression rate-distortion optimization (PCRD-opt).

The overall bit stream is built by concatenating the contributions from the independent code-block bit streams. The simplest concatenation organization

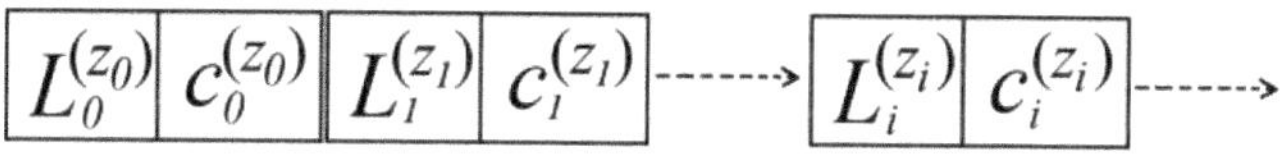

$$\boxed{L_0^{(z_0)}\ \ C_0^{(z_0)}}\ \boxed{L_1^{(z_1)}\ \ C_1^{(z_1)}}\ \text{------}\!\!\!\to\ \boxed{L_i^{(z_i)}\ \ C_i^{(z_i)}}\ \text{------}\!\!\!\to$$

Figure 8.16 Simple concatenation of optimally truncated code-block bit stream.

is illustrated in Figure 8.16; the length tags identify the contribution from each code-block.

Concatenation shown in Figure 8.16 is resolution and spatially scalable, but it is not distortion-scalable because each individual pack does not offer information to assist in the construction of a smaller stream whose code-block contributions are optimized in any way. However, EBCOT algorithm introduces the quality layer abstraction shown in Figure 8.17. The quality layers are labeled $Q_0 \cdots Q_{\Lambda-1}$. Each layer contains optimized code-block contributions of lengths L_i.

For example, the layer Q_0 contains optimized code-block contributions, having lengths $L_i^{(z_i^0)}$, which minimize the distortion $D^0 = \sum_i D_i^{(z_i^0)}$ subject to a length constraint $\sum_i L_i^{(z_i^0)} \leq L_{max}^0$. Remaining layers Q_λ contain additional contributions from each code-block, having lengths $L_i^{(z_i^\lambda)} - L_i^{(z_i^{\lambda-1})}$, which minimize the distortion $D^\lambda = \sum_i D_i^{(z_i^\lambda)}$ subject to $\sum_i L_i^{(z_i^\lambda)} \leq L_{max}^\lambda$.

Each quality layer may or may not contain a contribution from every code-block; for example, in Figure 8.17, code-block B_3 makes no contribution to layer Q_1.

The bitplanes of each code-block are inspected; if the MSB of a quantized index is marked as significant, its sign is encoded and its remaining bits will be refinement bits subsequently. If a quantized coefficient is zero, it is called insignificant. Hence, the EBCOT algorithm is conceptually divided into two layers called Tiers.

Tier 1 is responsible for source modeling and entropy coding, while Tier 2 generates the output stream (packetization process and supporting syntax). JPEG 2000 uses an efficient coding method for exploiting the redundancy of the bitplanes known as context-based adaptive binary arithmetic coding (CABAC). The current or decision bit is entropy coded using an arithmetic coder whose probability estimates are updated using a finite automaton with 46 states.

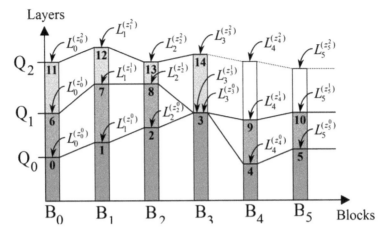

Figure 8.17   Quality layers. Numbers inside a layer indicate sequence of code-block contributions required for a quality progressive stream.

The probability estimate is updated separately for each of 18 different classes using separate contexts. Contexts 0–8 are used for significance coding during the significance propagation and clean-up passes, contexts 9–13 are used for sign coding, contexts 14–16 are used during the refinement pass, and an additional context is used for run coding during the clean-up pass.

Contexts are determined by coding primitives, neighborhood statistics, and subband frequency properties and refer to all the information needed to keep track of the current state of the finite automaton, which guides the statistical update of each coding class. In other words, symbol probabilities are estimated taking into account what has been encoded in the previous bitplanes and the neighboring bits. Adaptivity is provided by updating the probability estimate of a symbol based upon its present value and history.

Three different coding passes, referred to as sub-bitplanes, are used: significance propagation, magnitude refinement, and normalization (clean-up). The scan pattern followed for the coding of bitplanes, within each code-block (in all subbands), is shown in Figure 8.18. This scan pattern is basically a column-wise raster within stripes of height four. At the end of each stripe, scanning continues at the beginning (top-left) of the next stripe, until an entire bitplane (of a code-block) has been scanned. The prescribed scan is followed in each of the three coding passes.

During the significance propagation pass, the insignificant coefficients that have the highest probability of becoming significant, as determined by their immediate eight neighbors, are encoded. A coefficient is significant

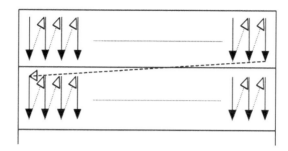

Figure 8.18   Scan pattern for bitplane coding within a code-block.

if there is at least one non-null bit among its corresponding bits coded in previous bitplanes.

In the magnitude refinement pass, the significant coefficients are refined by their bit representation in the current bitplane. A bit is coded in this pass if it belongs to an already significant coefficient. Significance propagation is used to propagate significance information. When a coefficient becomes significant, the probability that its neighboring coefficients become in turn significant grows higher and higher as the encoding proceeds.

During the normalization pass (clean-up pass), all bits belonging to non-significant coefficients having no significant coefficient in their neighborhood are coded in the normalization pass. Coefficients encoded in this pass have a very small magnitude and are expected to remain insignificant. Table 8.5 shows an example of the coding order for the quantized coefficients of one four-sample column in the scan. The example assumes that all neighbors not included in the table are identically zero and indicates in which pass each bit is coded. The very first pass in a new block is always a clean-up pass.

Table 8.5   Example of sub-bitplane coding order.

	Coefficient values			
Coding pass	12	1	5	−7
Clean-up	1+	0	0	0
Significance			1+	
Refinement	1	0		
Clean-up				1−
Significance				
Refinement	0	0	0	1
Clean-up				
Significance		1+		
Refinement	0		1	1
Clean-up				

The symbol to be encoded and the relative context are passed to the arithmetic coder called MQ-Coder. Then, the state is updated in order to refine probability estimates for the current context. The MQ-Coder encodes a binary information which indicates if the symbol being coded is a most probable symbol (MPS) or a less probable symbol (LPS). All contexts are initialized to uniform probabilities except for the zero context, where all neighbors are insignificant, and the run context, where the initial LPS probabilities are set to 0.030053 and 0.063012, respectively.

JPEG 2000 allows for the termination of the arithmetic-encoded bit stream as well as the reinitialization of the context probabilities at each coding pass boundary to decouple the arithmetic encoding of the sub-bitplane passes from one another, and allow the parallel encoding or decoding of the sub-bitplane passes of a single code-block (Taubman & Marcellin, 2001). A second option is the vertically stripe-causal contexts that enable the parallel decoding of the coding passes as well as reducing the external memory utilization. Another entropy coding option is the lazy coding mode, where the arithmetic encoder is entirely bypassed in certain coding passes. In this mode, it is necessary to terminate the arithmetic encoder at the end of the clean-up pass preceding each raw coding pass and to pad the raw coding pass data to align it with the byte boundary.

8.11 Rate Control

The rate control allows to specify a limit on image size during the encoding process. JPEG 2000 uses post-compression rate-distortion optimization (PCRD-opt) scheme in two steps to control the bit rate. In the first step, the distortion contribution and the bit rate of each coding pass is used to calculate the set of feasible truncation points of code-blocks.

The second step uses the generalized optimal Lagrange multiplier to estimate the set of truncation points for a target bit rate. This process approximates the optimal solution and has a low computational complexity. Quality layers allow to specify multiple image size limits and thus provide the possibility of progressive image quality improvement especially useful in a networked environment.

Overcoding occurs when more coding passes are generated for a code-block than will eventually be included in the final stream. If the quantizer step size is chosen to be small enough, the R-D performance of the algorithm is independent of the initial choice of the step size. Then, a Lagrangian R-D optimization determines the number of coding passes from each code-block that should be included in the final compressed bit stream to achieve the desired bit rate. If more than a single layer is desired, this process can be repeated at the end of each layer to determine the additional number of coding passes from each code-block that needs to be included in the next layer.

The wavelet coefficients $y(u,v)$ contained in a code-block of subband b are initially quantized with a step size of Δ_b, resulting in an M_b-bit quantizer index for each coefficient; (u,v) are the row and column indices. If the code-block bit stream is truncated so that only N_b bits are decoded, the effective quantizer step size for the coefficients is $\Delta_b 2^{(M_b-N_b)}$. The inclusion of each additional bitplane in the compressed bit stream will decrease the effective quantizer step size by a factor of 2. The effective quantizer step size might not be the same for every coefficient in a given code-block because of the inclusion of some coefficients in the sub-bitplane at which the truncation occurs.

Suppose that the entire image has a total number of code-blocks P and each code-block is denoted by $y(u,v)$. For a given truncation point t in code-block B_i, the weighted MSE distortion D_i^t is given by

$$D_i^t = \alpha_b^2 \sum_{u,v} w_i(u,v)\left[y_i(u,v) - y_i^t(u,v)\right]^2 \qquad (8.52)$$

where $w_i(u,v)$ is a weighting factor, α_b is the L_2-norm for subband b, and $y_i(u,v)$ and $y_i^t(u,v)$ are the original and the quantized coefficient values for truncation point t. If the $w_i(u,v) = 1$ for all subband coefficients, the distortion metric reduces to the MSE. At a given truncation point t, the rate for the code-block B_i is denoted by R_i^t. Given a total bit budget of R bytes, the rate control algorithm finds the truncation point for each code-block that minimizes the total distortion D. If we denote the optimal bit allocation as $\{R_i^* : 1 \le i \le P\}$, then the following linear programming program can be solved:

$$D = \sum_i D_i^*$$

$$\text{s.t.} \sum_i R_i^* \le R. \qquad (8.53)$$

For the weighting factor, several weighting strategies such as fixed visual weighting, bit-stream truncation and layer construction and progressive visual weighting can be used in conjunction with the EBCOT rate control algorithm (Taubman, Ordentlich, Weinberger, & Seroussi, 2002).

8.12 Region of Interest (ROI) Coding

Some applications require certain portion of the image with higher quality relative to the rest of the image, this portion is called region of interest or ROI. In this feature, the wavelet coefficients corresponding to the ROI are encoded by scaling up the ROI or scaling down the background of the ROI. The ROI is extracted using a binary mask called ROI mask before entropy coding. The bitplanes of coefficients belonging to the ROI mask are shifted up by a desired amount. The ROI coefficients are prioritized in the codestream so that they are received (decoded) before the background.

JPEG 2000 Part 1 adopted the Maxshift method to generate the ROI (Christopoulos, Askelof, & Larsson, 2000). The ROI-mask shape and the scaling factor are encoded and sent to the bit stream. The Part 2 ROI extension consists of Ericsson's scaling-based ROI method (Christopoulos, Askelof, & Larsson, 2000) and allows a single region of interest (RGN) marker (used to specify a region of interest) segment to signal an ROI covering all components. Therefore, some background can be included before ROI reaches the full accuracy and different priority levels can be selected.

8.13 Error Resilience

JPEG 2000 standard provides error resilience tools to deliver compressed data over distinct communication channels. The error resilience tools provided are such as compressed data partitioning and resynchronization, error detection, and quality of service (QoS) transmission based on priority. JPEG 2000 provides for the insertion of error resilience segmentation symbols ("1010" signaled in the end of marker segments encoded with the uniform arithmetic coding context) at the end of the clean-up pass of each bitplane that can serve in error detection. Error resilience at the packet level can be achieved by using the Start of Packet (SOP) marker segment, which provides for spatial partitioning and resynchronization.

8.14 JPEG 2000 Part 2

JPEG 2000 Part 2 is a collection of options implemented for specific market requirements. The major topics and locations are listed in the Table 8.6 following.

Table 8.6 Major topics of JPEG 2000 Part 2.

Compression efficiency	Annex	Functionality	Annex
Variable DC offset (VDCO)	B	Geometric manipulation	I
Variable scalar quantization (VSQ)	C	Single-sample overlap (SSO/TSSO)	I
Trellis coded quantization (TCQ)	D	Precinct-dependent quantization	Amendment 1
Extended visual masking	E	Extended region of interest	L
Arbitrary wavelet decomposition	F	Extended file format/metadata (JPX)	M,N
Arbitrary wavelet transform kernel	G, H	Extended capabilities signaling	Amendment 2
Multiple component transform	J		
Nonlinear point transform	K		

8.15 Motion JPEG 2000

Motion JPEG 2000 (ITU-T-T.809, 2011) is the video version of JPEG 2000 in which each frame is compressed frame by frame, independently of the motion information. Motion JPEG 2000 is defined in Part 3 of (ITU-T-T.809, 2011). This is the standard for the digital cinema format for digital movie theaters.

Motion JPEG 2000 (MJ2) is one potential format for long-term video preservation. The format is attractive as an open standard with a truly lossless compression mode. Some software for MJ2 implementations are readily available, from the Open JPEG 2000 project (UCL, 2019), from the Kakadu project (Warmerdam & Rouault, 2019), and (incorporating Kakadu) from vendor Morgan Multimedia. All the implementations have practical limitations on acceptable input formats and inadequate or missing audio support.

8.16 Volumetric Data Encoding

Volumetric data encoding is defined in JPEG 2000 Part 10 (ITU-T-T.809, 2011). This extension for three-dimensional data (JP3D) supports functionalities like resolution scalability, quality scalability, and region-of-interest coding; while exploiting the entropy in the additional third dimension to improve the rate-distortion performance (Schelkens, Munteanu, Tzannes, & Brilawn, 2006), JP3D supports existing capabilities of Part 1, extended to three dimensions. It also respects syntax and capabilities of Part 2 multi-component images, although it does not extend them all. Features relevant to encoding volumetric dataset, such as the arbitrary transformation kernels (ATK) and ROI coding, are extended, while less relevant features are not adopted.

JP3D packet syntax is compatible with the interactive protocols defined in JPIP4 (Part 9) for client–server interaction. This protocol enables applications to randomly access remote codestreams, thus exploiting the full scalability potentials of JPEG 2000 (Bruylants, Munteanu, Alecu, Decklerk, & Schelkens, 2007).

The JP3D codec first divides the input data in volumetric tiles as shown in Figure 8.19. Tiles in this case are cuboid subvolumes that are independently encoded. All components are encoded as separate gray-scale datasets. Next, the DWT filters all tile sample coefficients with number of decomposition levels in each of the three dimensions freely chosen (Das, Hazra, & Banerjee, 2010), (Biswas, Malreddy, & Banerjee, 2018).

Each cubic subband is split into small dyadically sized cubes, called code-blocks, and sent to the EBCOT coder to be independently encoded.

The block-scanning pattern is shown in Figure 8.20 where the 3D code-block coefficients are processed in a bitplane-per-bitplane and slice-by-slice order as in aJ PEG 2000 Part 1 pattern. Bitplanes are processed from the most significant bitplane to the least significant bitplane, while, within each bitplane, each "slice" is scanned in turn. In each slice, the scan order is the same as in a 2D block.

Once a bitplane of a slice is completely processed, the algorithm continues with next slice in the code-block.

JP3D specification uses EBCOT, based on the principles of layered zero coding (LZC), which is the core element of the JPEG2000 image compression standard.

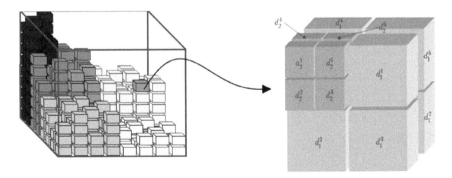

Figure 8.19 Partitioning of volumetric data in volumetric tiles (left) and subband decomposition (right) (Bruylants, Munteanu, Alecu, Decklerk, & Schelkens, 2007).

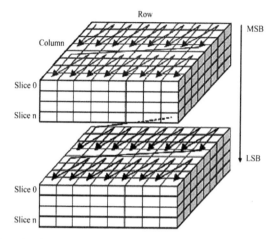

Figure 8.20 3D code-block scanning pattern (Bruylants, Munteanu, Alecu, Decklerk, & Schelkens, 2007) showing the scanning of two 3D 8 × 8 × 4 code-block consisting of four 8 × 8 slices.

8.17 Some Limitations of JPEG 2000

JPEG 2000 standard exhibits undesirable limitations that avoid its universal use and support; some of these limitations are listed as follows:

- Encoding JPEG 2000 is not as fast and easy as encoding JPEG.
- Choosing a very low bit rate will result in a mess of an image. The target bit rate will have to be adjusted manually depending on the content of the image.
- As in the JPEG case, in JPEG 2000, iterating compression over multiple processing stations, i.e., an in-house intranet cabling will pile up latency and iterative re-compression will introduce additional compression defects (Richter, Keinert, Descampe, & Rouvroy, 2018).
- Images are therefore separated into tiles that are then compressed one by another. This degrades image quality and is not necessarily ideal in terms of rate distortion; however, WG1 found that JPEG 2000 alone, and even one of the proposed modifications of it, tentatively named FBCOT (Taubman, Naman, & Mathew, 2017), would probably not fit into the low-cost requirements.
- Websites and camera manufacturers were not ready to accept the format until it was widely adopted.
- The format is not backward compatible with JPEG standard.
- No universal browser support. Users are increasingly using browsers to perform tasks that used to be considered solely the domain of traditional desktop applications.

• Complexity of JPEG 2000 significantly increases the barrier to entry for new implementations.

8.18 Summary

JPEG 2000 is an advanced image compression standard, which incorporates and emphasizes many features not found in earlier compression standards. Many of these features are imparted by independent, embedded coding of individual blocks of subband samples.

JPEG 2000, which is largely derived from EBCOT, is based on the DWT, scalar quantization, context modeling, arithmetic coding, and post-compression rate allocation. The entropy coding is done in blocks, typically 64×64, inside each subband. The DWT can be performed with reversible filters, which provide for lossless coding, or non-reversible filters, which provide for higher coding efficiency without the possibility to do lossless.

The coded data is organized in so-called layers, which are quality levels, using post-compression rate allocation and then output to the codestream in packets. JPEG 2000 provides for resolution, SNR and position progressivity, or any combination of them, parsable codestreams, error-resilience, arbitrarily shaped region of interest, random access (to the subband block level), lossy and lossless coding, etc., all in a unified algorithm.

Part 1.   Includes limited ROI capability provided by the Maxshift method. The Maxshift method has the advantage that it does not require the transmission of the ROI shape and it can accommodate arbitrarily shaped ROIs. On the other hand, it cannot arbitrarily control the quality of each ROI with respect to the background.

Part 2.   Extends the ROI capability by allowing the wavelet coefficients corresponding to a given ROI to be scaled by an arbitrary scaling factor. However, this necessitates the sending of the ROI information explicitly to the decoder, which adds to the decoder complexity. Part 2 supports only rectangular and elliptic ROIs, and the ROI mask is constructed at the decoder based on the shape information included in the codestream.

Part 2 of the JPEG 2000 standard offers two ways of nonlinear transformation of the component samples before any inter-component transform is applied to increase coding efficiency. The two nonlinear transformations are gamma-style and lookup table (LUT) style. This feature is especially useful when the image components are in the linear intensity domain, but it is desirable to bring them to a perceptually uniform domain for compression that is more efficient. For example, the output of a 12-bit linear sensor or scanner can be transformed to 8 bits using a gamma or a logarithmic function.

JPEG 2000 provides, in most cases, competitive compression ratios with the added benefit of scalability and higher protection than JPEG. However, JPEG 2000 standard is substantially more complex than the baseline sequential JPEG algorithm, both from a conceptual and an implementation standpoint (Rabbani & Joshi, 2002), (Santa-Cruz, Grosbois, & Ebrahimi, 2002).

JP3D is a component of the JPEG 2000 suite. It is a straightforward extension of JPEG 2000 Parts 1 and 2, designed for volumetric encoding to support lossy-to-lossless coding functionality (quality scalability), resolution scalability, and region-of-interest coding (Zhang, Fowler, Younan, & Liu, 2009).

9

JPEG XS

The JPEG XS (ISO/IEC-DIS-21122, Under development), (JPEG, Overview of JPEG XS, 2018) standard defines a compression algorithm with very low latency and very low complexity, particularly optimized for visual lossless compression (ISO/IEC-29170-2, 2015) for both natural and synthetic images.

By offering various degrees of parallelism, JPEG XS can be efficiently implemented on various platforms such as FPGAs, ASICs, CPUs, and GPUs and excels with high multi-generation robustness (JPEG, Overview of JPEG XS, 2018). The typical compression ratios are between 1:2 and 1:6 for both 4:4:4 and 4:2:2 images and image sequences with up to 16-bit component precision. Typical parameterizations address a maximum algorithmic latency between 1 and 32 video lines for a combined encoder–decoder suite. The main goal of JPEG XS is to enable cost-efficient, parallel implementations in FPGAs of GPUs rather than obtaining an optimal rate-distortion performance (Richter, Keinert, Descampe, & Rouvroy, 2018).

This codec is intended to be used in applications where content would usually be transmitted or stored in uncompressed form such as in live production, display links, virtual and augmented reality, self-driving vehicles, or frame buffers (Willème *et al.*, 2018).

9.1 JPEG XS Parts

JPEG XS is a multi-part specification being currently under development and includes the following five parts:

Part 1. Core coding system: JPEG XS Part 1 (ISO/IEC 21122-1) normatively defines how a compressed JPEG XS codestream can be transformed into a decoded image in a bit exact manner. Moreover, it informatively explains the key algorithms enabling an encoder to generate a JPEG XS codestream.

Part 2. Profiles and buffer models: JPEG XS Part 2 (ISO/IEC 21122-2) ensures interoperability between different implementations by specifying typical codestream parameterizations and properties. This allows deriving the hardware and software requirements for different purposes such as high compression ratios, low memory, or low logic implementations. Moreover, implementation guidelines inform about how to achieve low latency implementations.

Part 3.   Transport and container: JPEG XS Part 3 (ISO/IEC 21122-3) defines how to embed a JPEG XS codestream into a more descriptive file format. Moreover, it contains all definitions that are necessary to transport a JPEG XS codestream by means of a transmission channel using existing transmission protocols defined by different standardization bodies.

Part 4.   Conformance testing: JPEG XS Part 4 defines conformance testing of JPEG XS.

Part 5.   Reference software: JPEG XS Part 5 provides the JPEG XS reference software.

JPEG XS will offer the following features (JPEG, 2018).

- Safeguarding all advantages of an uncompressed stream:
  - low power consumption (through lightweight image processing);
  - low-latency in coding and decoding – easy to implement (through low complexity algorithm) – small size on chip and fast software running on general purpose CPU with the use of SIMD and GPU.
- Without significant increase in required bandwidth:
  - low power consumption (through reasonable bandwidth interfaces);
  - longer cable runs;
  - SRAM size and frequency reduction with a frame buffer compression;
  - more adequate for current infrastructures.

9.2 Architecture of JPEG XS

Figure 9.1 shows the block diagram of the architecture of the JPEG XS (a) encoder and (b) decoder. The input components are decorrelated using a lossless color transform identical to the one used in JPEG 2000. An integer irreversible DWT is applied to each component. The resulting wavelet coefficients are analyzed by a budget to predict the number of bits required for each possible quantization. The rate control algorithm computes the smallest quantization factor that does not exceed the bit budget available for coding the wavelet coefficients. Then the wavelet coefficients are entropy coded. Finally, all data sections are combined into a packet structure and sent to the transmission channel.

Given that the decoder should be able to process the pixels with a constant clock frequency, the number of bits read per time unit varies depending on

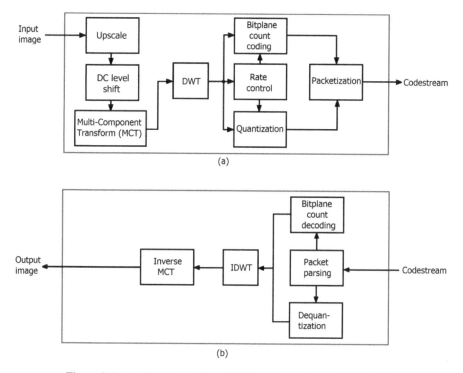

Figure 9.1   Architecture of the JPEG XS (a) encoder and (b) decoder.

whether a current wavelet coefficient is easy to compress or not. A packet parser splits the bit stream into individual data chunks representing parts of a subband before the wavelet coefficients are decoded and inverse transformed to recover the image.

9.3 Multi-Component Transform

The input image is up-scaled to 20-bit precision and the DC offset is removed to create a symmetric signal around zero. For RGB input, a color transformation into a YCbCr color space identical to RCT used in the JPEG 2000 is applied.

9.4 Forward Transform

The specifications allow a one- or two-level vertical transformation with the LeGall 5/3 wavelet (see Section 8.7.8) and a five-level horizontal transformation. Periodic symmetric extension is used and the number of additional samples required at the boundaries is filter length dependent.

9.5 Pre-Quantization

The wavelet stage is followed by a pre-quantizer that first represents data as sign and magnitude and then chops off the 8 least significant bitplanes. This is equivalent to a dead-zone quantizer with a quantization bucket size of $\Delta = 256$ and a dead-zone whose size is twice the size of a regular bucket. The pre-quantization step is not used for rate-control purposes and only ensures that the following data path can be 16 bits wide (Richter, Keinert, Descampe, & Rouvroy, 2018).

9.6 Quantization

Rate control is performed via quantization step. Besides a dead-zone quantizer similar to JPEG 2000, JPEG XS uses a data-dependent uniform quantizer that operates on the pre-quantized wavelet coefficients values clustered into groups of four coding groups. For each coding group, the M_g quantity is defined as the number of non-zero bitplanes of the coding group called also the MSB position

$$M_g = \max\left(\left\lfloor \log_2 \max_{i \in g}(x_i) \right\rfloor + 1, \ 0\right) \tag{9.1}$$

where g is the coding group and x_i is the ith coefficient inside the coding group g.

When the dead-zone quantizer is selected, the full interval of values $\left(-2^{M_g}, \ 2^{M_g}\right)$ in a coding group is divided by the quantization step size $\Delta = 2^T$, leaving a total of $2^{M_g+1-T} - 1$ including the zero bucket (zero bucket extends from -2^T to 2^T which is twice the size of a regular bucket).

When the uniform quantizer is selected, the bucket size is

$$\Delta_T = \frac{2^{M_g+1}}{2^{M_g+1-T} - 1} = 2^T \left(\frac{2^{M_g+1-T}}{2^{M_g+1-T} - 1}\right) \tag{9.2}$$

which is the same output range and the same number of buckets as the dead-zone quantizer. Note that Equation (9.2) is easy to implement at bit level; first left-shift by $M_g+1 - T$ bits and then subtract and right-shift by M_g+1 bits. Reconstruction is more complex because it requires multiplication by Δ_T and a division by a number which is not a power of 2. However, we can express Equation (9.2) using Newman series as

$$\Delta_T = 2^T \left(\frac{\lambda}{\lambda - 1}\right) = 2^T \left[\frac{1}{1 - \frac{1}{\lambda}}\right] = 2^T \left(\frac{1}{1 - \lambda^{-1}}\right) = 2^T \sum_{k=0}^{\infty} \lambda^{-k} \tag{9.3}$$

where $\lambda = 2^{Mg+1-T}$. Reconstruction with Equation (9.3) implies that the quantized coefficients c be multiplied by Δ_T. This operation can be carried out by first left-shifting the coefficient T bits (multiplication by 2^T) and then summing up right-shifted copies of $c \times 2^T$ by $M_g + 1 - T$ bits. The infinite summation can be truncated by checking when all significant digits are shifted out. The operation can be expressed as $\lfloor c \times \Delta_T \rfloor$, where $\lfloor \cdot \rfloor$ is the floor operation.

9.7 Rate Control

JPEG XS divides into rectangular regions or precincts p, the wavelet coefficients, similar to JPEG 2000. Nevertheless, JPEG XS precincts include only one or two lines (compared to thin rectangular precincts in JPEG 2000 where the dimension is scaled with 2^r, with r being the resolution level). Rate allocation assigns a quantization parameter $T_{b,p}$ to each band b in each precinct p, except that the rate allocator of JPEG XS is – due to latency constraints – unable to operate on a full frame but has only access to a limited window of the image. Hence, it is a heuristics and not a precise algorithm (Richter, Keinert, Descampe, & Rouvroy, 2018).

Two parameters are defined per precinct, the quantization Q_p and the refinement R_p. Two additional parameters per band (which are properties of the wavelet filters and stay constant throughout the frame) are defined, the band gains Q_b and the band priorities P_b. The $T_{b,p}$ parameter can be derived as

$$T_{b,p} = Q_p - G_b + r, \qquad r = 1 \text{ for } P_b < R_p \text{ and } 0 \text{ otherwise.}$$

Q_b and P_b depend only on the wavelet filter and thus only require signaling once per frame, instead of transmitting the full set of $T_{b,p}$ separately for each precinct.

For ideal MSE performance (not necessarily for ideal rate-distortion performance), the quantization of band b should be derived from a base quantization Δ_T as

$$\Delta_b = \frac{\Delta_0}{\gamma_b} \qquad (9.4)$$

where $\gamma_b := \left\| W_b^{-1} \delta \right\|_2$, W_b^{-1} is the inverse wavelet filter of band b, $\left\| \cdot \right\|_2$ is the ℓ_2 norm (square root of the mean square error), and δ is the impulse signal. This is the ℓ_2 error contribution of wavelet band b in the image domain. The approximation holds exactly when DW coefficients are statistically decorrelated (Richter, Keinert, Descampe, & Rouvroy, 2018).

To reduce complexity, JPEG XS only attempts to reach the target rate by modulating the band quantization parameters $T_{b,p}$ without measuring the distortion. Note that

$$T_{p,b}^0 \approx \left\lceil \log_2 \left(\frac{\Delta_p}{\gamma_b} \right) \right\rceil \approx \left\lceil \log_2 \Delta_p \right\rceil - \left\lceil \log_2 \gamma_p \right\rceil = Q_b - G_b \qquad (9.5)$$

where Δ_p is the precinct (but not band)-dependent base-quantization into which we absorb any proportionality factors and $\lceil \cdot \rceil$ is the ceil operation. Observe that $\left\lceil \log_2 \gamma_p \right\rceil = G_b$; then we can define the factor

$$\mu_b := \gamma_b / 2^{G_b}. \qquad (9.6)$$

Priorities P_b are assigned by ordering bands in order of decreasing μ_b. For a given set of $T_{p,b}^0$, additional rate remains available, and the excess rate can be made use of by including one additional bitplane in the order given by μ_b. The rate-allocator should consider those bands first whose rank in terms of μ_b is lowest or whose μ_b is the highest. Signaling the rank order P_b of bands in the μ_b list per frame and rank R_p up to which rank bands are refined is hence sufficient. The precise way as to how a rate allocator operates is not specified in the standard and multiple algorithms including heuristics from past frames or last lines can be considered.

9.8 Entropy Coding

Entropy coding in JPEG XS is, aligned with the low-complexity design goal, minimalistic: Once wavelet coefficients have been quantized, quadruples of wavelet coefficients are formed. These are the same groups that have already been set up for parameterizing the data-dependent quantizer. For each group, the following datasets are formed: First, the MSB positions of all groups, indicating how many bitplanes of quantized data, are included, second, the quantized coefficient values themselves, and, third, the signs of all non-zero coefficients in the group. Of all the three datasets, only the first one is entropy coded, and all the remaining ones are transmitted directly, without any coding.

Even though it seems a bit counter-intuitive, the first group – the MSB positions – requires a major part of the overall rate. Due to the energy compaction property of the wavelets, at typical operation points, typically no more than two bitplanes have to be included at all. Multiple methods of MSB position coding had been discussed within WG1, and we shall only describe a subset of everything that has been proposed.

All proposed methods are based on prediction coding where, optionally, the MSB position of the top neighbor of a coding group is used to predict the MSB position of the current coding group. On top of that, additional schemes are under discussion that signal long runs of zero MSB positions or zero MSB position predictions. Only intra-band prediction schemes are currently considered. Horizontal prediction had been removed because it does not provide competitive gains and is harder to parallelize.

One of the proposed prediction schemes operates as follows: Given the MSB position of the current group M_g, the MSB position of a past group M'_g and the truncation positions $T_{p,b}$ of the current and the past precinct $T_{p',b}$, then the prediction residual $\delta_{g,g'}$ is defined as (Richter, Keinert, Descampe, & Rouvroy, 2018)

$$\delta_{g,g'} = \max\left(M_g, T_{p,b}\right) - \max\left(\max\left(M_{g'}, T_{p',b}\right), T_{p,b}\right). \tag{9.7}$$

JPEG XS has a limited flexibility of architecture because it is very hardware-oriented. A very simple definition of a run is to combine consecutive sets of coding groups into run groups, e.g., groups of eight coding groups each, and identify a run group as insignificant if either all prediction residuals within the run group are zero or if $\max\left(M_{g'}, T_{p',b}\right) = T_{p,b}$, i.e., all coefficient data is truncated away.

According to Richter, Keinert, Descampe, and Rouvroy (2018), the residual is bounded from below by

$$\delta_{g,g'} \geq \Theta_{g',p',p,b} = \min(T_{p,b} - \max(M_{g'}, T_{p',b}), 0). \tag{9.8}$$

Also, note that

$$\left|\delta_{g,g'}\right| \geq \left|\Theta_{g',p',p,b}\right| = \max(\max(M_{g'}, T_{p',b}) - T_{p,b}, 0). \tag{9.9}$$

Therefore, as $\delta_{g,g'}$ can take positive and negative values, the use of the following Golomb–Rice is possible on the residual

$$\begin{cases} 1^{2|\delta_{g,g'}|-1}0 & \text{for } \delta_{g,g'} < 0 \\ 1^{2|\delta_{g,g'}|}0 & \text{for } \delta_{g,g'} \geq 0 \text{ and } \left|\delta_{g,g'}\right| \leq \left|\Theta_{g',p',p,b}\right| \end{cases}. \tag{9.10}$$

If $\left|\delta_{g,g'}\right| > \left|\Theta_{g',b,b}\right|$, negative code words are not possible and single-side code is sufficient; otherwise, code words are assigned to an interleaved sequence of positive and negative code words.. When $\left|\delta_{g,g'}\right| > \left|\Theta_{g',b,b}\right|$, the encoding is similar to the JPEG-LS (Weinberger, Seroussi, & Saprio)

$$1^{\delta_{g,g'}+\left|\Theta_{g',p',b}\right|}0 \quad \text{for } \delta_{g,g'} \geq 0 \text{ and } \left|\delta_{g,g'}\right| > \left|\Theta_{g',p',p,b}\right|. \tag{9.11}$$

Compression efficiency can be improved by encoding long runs with less symbols. The committee draft text specifies two definitions of what establishes a run, and, historically, multiple encoding mechanisms for signaling a run have been proposed, of which only one was adopted in the committee draft.

A run can be defined in two forms. The first combines consecutive sets of coding groups into run groups, e.g., groups of eight coding groups each, and identifies a run group as insignificant if either all prediction residuals within the run group are zero or if all coefficient data is truncated away, i.e., $\max\left(M_g, T_{p,b}\right) = T_{p,b}$.

The second definition of a run is detected at the output logic on the quantizer; in this case, $T_{p,b} > M_g$. In the first case, the rate saving for a run is N bits because, in the second case, the rate saving is less trivial. Then, the prediction residual can be expressed as

$$\delta_{g,g'} = T_{p,b} - \max(\max(M_{g'},T_{p',b}),T_{p,b}) = -\max(\max(M_{g'},T_{p',b})-T_{p,b},0). \tag{9.12}$$

Only negative values contribute to runs.

9.9 Subjective Evaluations and Setup

JPEG XS takes advantage of the contrast sensitivity function and the visual masking to allow near lossless and visual lossless coding. Several configurations based on the Draft International of JEPG XS Part 2 have been tested in JPEG XS for subjective assessment campaign (Willème *et al.*, 2018).

- Main configuration (related to Main profile) provides an intermediate tradeoff between complexity, buffering needs, and coding efficiency. It enables five horizontal DWT decompositions using a 5–3 wavelet structure but only one single vertical DWT decomposition. It allows for three different prediction modes for the bitplane counts. It targets use cases such as broadcast production, Pro-AV, frame buffer compression (FBC) in digital TV (DTV), and compression over display links.
- High configuration (related to High profile) aims at providing the best coding efficiency (and visual quality preservation) at the cost of increased complexity and more buffering. Compared to the Main configuration, it

enables two vertical DWT decompositions and uses additional coding tools within its entropy coder. Targeted applications include high-end Pro-AV, broadcast contribution, cinema remote production, FBC, and compression over display links in high-end DTV.

- Low logic configuration (related to Light profile) models applications that can only afford the lowest complexity implementations at the cost of a slight coding efficiency drop. Compared to the Main configuration, only two prediction modes for the bitplane counts are available and the quantizer is simpler. It is intended to be used in use cases such as broadcast production, industrial cameras, and in-camera compression (for compression ratios up to 4:1).
- Low memory configuration (related to Light-Subline profile) is intended at reducing significantly the buffering needs at the cost of a coding efficiency drop. To do so, it does not perform any vertical DWT decomposition. It also splits the frames into two columns that are encoded independently. Targeted use cases include in-camera compression (for compression ratios up to 2:1) and many cost-sensitive applications.

The evaluation included the following setup described in Annex B of ISO/IEC 29170-2 (AIC Part-2) (ISO/IEC-29170-2:2015, 2018):
- Monitor Eizo CG318-4k with 4K UHD resolution (3840×2160 pixels) and 10-bit color depth processing (EIZO-Corporation, 2018). Monitor settings were applied in order to minimize additional stimuli processing and the color representation mode was set to ITU-R BT.709-6 (ITU-R-BT.709-6, 2015).
- Observers were seated at the viewing distance of 60 ppd (pixels per degree) which approximately corresponds to 58.6 cm for Eizo CG318-4k (native resolution 3840×2160 pixels, viewable image size 698.0×368.1 mm).
- Head/chin-rest (recommended in ISO/IEC 29170-2) was not used, only one single observer was viewing the monitor at a time (the viewing distance could not be kept exact to the specification during the entire test session).
- Viewing room with controlled lighting similar or compliant with ITU-R BT.500-13.15 (ITU-R-BT.500-13, 2012). Background luminance was low in the range between 15 and 20 cd/m^2.
- Target peak screen luminance was set to 120 cd/m^2 with allowed luminance range of 100–140 cd/m^2 for hardware calibration of monitor.
- The resolution was set to 3840×2160, refresh rate 24 Hz, and 10 bpc output activated.

- Open source and cross-platform media player with the proper setting were capable of correctly displaying stimuli with 10 bpc color content (MPV, 2018).
- All the test laboratories used the same testing software package including scripts for evaluation sessions and tools for automatic consolidation of obtained results.

The following empirical metric was used to obtain a score for each test image:

$$\text{Score} = 2 \times \left(1 - \frac{OK + 0.5 \times ND}{N} \right), \tag{9.13}$$

where N is the total number of votes for a given stimulus, OK is the number of times the flickering is correctly identified, and ND is the number of times that a subject decided to cast a no-decision vote. The score is bounded to [0, 2], where 0 means all subjective votes are correct; score 1 is achieved when all the votes are casted as no decision, implying that no subject could identify flickering (transparent coding). The upper bound score 2 is obtained if all votes are wrong. This case is avoided by using the outlier removal procedure described next. Subjects who either misunderstood the test procedure or who appeared to have delivered insufficient effort in spite of the given instructions were removed. To remove unreliable votes, the following threshold was applied to exclude subjects with too many wrong decisions:

$$\text{Threshold} = \mu_{\text{Wrong}} + 2 \times \sigma_{\text{Wrong}}, \tag{9.14}$$

where μ_{wrong} and σ_{wrong} are the mean and standard deviation of the number of "Wrong" decisions per subject. Subjects with a number of "Wrong" decisions larger than the threshold were excluded from the test.

During a session, each stimulus (one image encoded by one JPEG XS configuration at one given bit rate) was shown 8 times at randomized points (instead of 30 times as defined in AIC Part-2) during a maximum of 5 seconds. In order to improve the reliability of the assessment, both training stimuli and control stimuli were inserted in the test material.

Training stimuli use encoded–decoded images that present obvious compression artifacts. They were shown to the subject at the beginning of the session, with an instructor beside, to help identify visual discrepancies. The encoded–decoded images were undergone seven successive encoding–decoding cycles. This allows one to assess the multi-generation robustness of JPEG XS. The bit rates used for encoding images varied from one image to another and also depended on the JPEG XS configuration.

The evaluation was conducted in parallel by three test laboratories affiliated to, respectively, Czech Technical University in Prague (CTU), Ecole Polytechnique Fédérale de Lausanne (EPFL), and Vrije Universiteit Brussel-imec (VUB-imec).

9.10 Test Material

9.10.1 VQEG

Sequences S01, S02:
Individuals and organizations extracting sequences from this archive agree that the sequences and all intellectual property rights therein remain the property of Sveriges Television AB (SVT), Sweden. These sequences may only be used for the purpose of developing, testing, and presenting technology standards. SVT makes no warranties with respect to the materials and expressly disclaims any warranties regarding their fitness for any purpose.

9.10.2 ARRI

Sequences S03, S04:
These sequences have been provided by ARRI – Arnold and Richter Cine Technik GmbH Image Science Group, R&D Department Türkenstraße 89, 80799 Munich, Germany. The originals can be downloaded from ARRI (2019). This sample footage was created for workflow evaluation purposes. On Wednesday, 2016/03/09, Mr. Johannes Steurer from ARRI kindly granted WG1 the possibility to provide the converted material for download on a password protected FTP server, whereas the password needs to be queried from a JPEG representative.

Sequences S05, S06:
These sequences have been made public by ARRI – Arnold and Richter Cine Technik GmbH Image Science Group, R&D Department Türkenstraße 89, 80799 Munich, Germany (ARRI, 2019). The work is licensed under the Creative Commons Attribution 3.0 Unported License.

9.10.3 Blender

Sequences S07, S08 (Sintel):
© copyright Blender Foundation|durian.blender.org
   The results of the Durian Open Movie project are being licensed under the Creative Commons Attribution 3.0 license. In short, this means you

can freely reuse and distribute this content, also commercially, for as long you provide a proper attribution. See Blender Foundation (2019) for more info.

Sequence S09 (Tears of steel):
© copyright Blender Foundation|mango.blender.org
   The proceedings and results of the Mango Open Movie project are being licensed under the Creative Commons Attribution 3.0 license. In short, this means you can freely reuse and distribute this content, also commercially, for as long you provide a proper attribution. See Blender Foundation (2019) for more info.

9.10.4 EBU

Sequence S12 (PendulusWide)
Sequence S13 (RainFruits)

9.10.5 Apple

Sequence S14:
This sequence and all intellectual property rights therein remain the property of Apple, Inc. This sequence may only be used for the purpose of developing, testing, and promulgating technology standards. Apple, Inc. makes no warranties with respect to the sequences and expressly disclaims any warranties regarding their fitness for any purpose.
   The Basketball Drill sequence that appears in this sequence and all intellectual property rights therein remain the property of NTT DOCOMO, INC.

9.10.6 Huawei

Sequence S15:
The test sequence and all intellectual property rights therein remain the property of their respective copyright holders. This material can only be used for the purpose of academic research and development of standards. This material cannot be distributed with charge. The owner makes no warranties with respect to the material and expressly disclaims any warranties regarding its fitness for any purpose. Owners: Huawei Technologies and MERL. Production: Huawei Technologies and MERL.

9.10.7 Richter

Sequence S16:

The material in this directory is copyrighted (c) 2016 Thomas Richter. You are hereby granted to use this material under the Creative Commons 4.0 BY-SA license. Briefly, this means that you need to quote the author and are obliged to distribute adapted material based on this material under the same license conditions.

9.10.8 Copyright Condition Statements for Test Still Images

Zoneplates images:

2016 February 26 Images in the set prefix "Zoneplate_..." are copyright 2015-16 by Samsung Semiconductor, Inc.

Permission is given to use these images for coding system (compression) research and development.

Publication of a thumbnail is permitted in technical journals, and no rights to commercial use or advertisement at trade shows or other commercial venues is given. Contact: Dale Stolitzka, Samsung d.stolitzka@samsung.com.

HorseFly:

"Female Striped Horse Fly (Tabanus lineola)" by Thomas Shahan. This file is licensed under the Creative Commons Attribution 2.0 Generic license.

You are free:
• to share – to copy, distribute, and transmit the work;
• to remix – to adapt the work.

Under the following conditions:
• attribution – you must attribute the work in the manner specified by the author or licensor (but not in any way that suggests that they endorse you or your use of the work).

Source (accessed March 10, 2016):

https://commons.wikimedia.org/wiki/File:Female_Striped_Horse_Fly_ (Tabanus_lineola).png

9.10.9 Tools

File provided copyright free by Richard Clark/Crosfield. These pictures were taken on a visit to Crosfield by Richard Clark – Crosfield and he make no claims on copyright.

Peacock:

CC0 Public Domain. Free for commercial use. No attribution required
Source (accessed March 10, 2016):
https://pixabay.com/en/peacock-bird-feather-close-color-90051/

HintergrundMusik:

Background Music 203 by Sabine Sauermaul. License: Public Domain Dedication (You can copy, modify, distribute, and perform the work, even for commercial purposes, all without asking permission).
Source (accessed March 10, 2016):
http://www.publicdomainpictures.net/

LeavesIso200:

These images are available without any prohibitive copyright restrictions. These images are their respective owners. You are granted full redistribution and publication rights on these images provided:

1. The origin of the pictures must not be misrepresented; you must not claim that you took the original pictures. If you use, publish, or redistribute them, an acknowledgment would be appreciated but is not required.
2. Altered versions must be plainly marked as such and must not be misinterpreted as being the originals.
3. No payment is required for distribution of this material; it must be available freely under the conditions stated here. That is, it is prohibited to sell the material.
4. This notice may not be removed or altered from any distribution.

9.10.10 Acknowledgments

A lot of people contributed a lot of time and effort in making this test set possible. Thanks to everyone who shared their opinion in any of the discussions online or by email. Thanks to Axel Becker, Thomas Richter, and Niels Fröhling for their extensive help in picking images, running all the various tests, etc. Thanks to Pete Fraser, Tony Story, Wayne J. Cosshall, David Coffin, Bruce Lindbloom, rawsamples.ch, and raw.fotosite.pl for the images that make up this set.

Source (accessed March 10, 2016):
leaves_iso_200.ppm on http://imagecompression.info/test_images/

10

JPEG Pleno

JPEG Pleno (JPEG, Overview of JPEG Pleno, 2019) aims to provide a standard framework for representing new imaging modalities, such as texture-plus-depth, light field, point cloud, and holographic imaging. This goal requires the specification of system tools, coding tools, and appropriate metadata, not only for efficient compression but also for data and metadata manipulation, editing, random access and interaction, protection of privacy and ownership rights, as well as security management. This framework should offer more than the sum of its parts by allowing them to be flexibly combined, exchanged, and exploited in a single scene or processing pipeline.

JPEG Pleno standard tools will be designed together to consider their synergies and dependencies for the whole to be effectively greater than the sum of its parts. To fully exploit this holistic approach, JPEG Pleno is not just a set of efficient coding tools addressing compression efficiency. It is a representation framework understood as a fully integrated system for providing advanced functionality support for image manipulation, metadata, random access and interaction, and various file formats. In addition, it should offer privacy protection, ownership rights, and security.

The JPEG Pleno framework goes end-to-end from the real or synthetized world to the replicated world, harmoniously integrating all necessary tools into a single system to represent the same visual reality while considering different modalities, requirements, and functionalities (JPEG, Overview of JPEG Pleno, 2019), (JPEG, Workplan & Specs of JPEG Pleno, 2018), (JPEG, Documentation on JPEG Pleno, 2018), (JPEG, JPEG Pleno Database, 2018).

10.1 Plenoptic Representation

Images and video have been traditionally acquired, represented, coded, and displayed using a simple model corresponding to rectangular sets of samples for some wavelength components, e.g., RGB or luminance and chrominance. However, the interest of more immersive experiences with richer and more complete representation of the visual information has increased; thus, it is precise to include more useful models to represent the reality more accurately. For example, a model that includes the description on how the waves of different energy (power density) varying in time *t* (dynamic scenes), at a rate much smaller than the period of the sinusoidal electromagnetic waves and a spatial position (x,y,z).

The wave propagates as a sum of random-phase sinusoids within the visible spectrum. Each wave has a different wavelength λ. The waves can be decomposed by a sum of wave fronts coming from all directions with certain azimuth (θ) and orientation (Φ). Hence, a seven-dimensional (7D) scalar function $P(x,y,z,\theta,\Phi,\lambda,t)$ can be used to symbolize a plenoptic function, provided the light is incoherent. The plenoptic function assumes a volume through which all the light rays flow and represents an enormous amount of data; Figure 10.1 shows the parameterization of one ray in the 3D plane by position (x,y,z), azimuth, and elevation (θ,Φ). Therefore, it is essential to reduce its dimensionality and appropriately sample the plenoptic function (Pereira & da Silva, 2016), and, at the same time, sampling of the plenoptic function dimensions has to be such that aliasing can be avoided (Do, M-Maillet, & Vetterli, 2012), (Gillian, Dragotti, & Brookes, 2014).

Imagine that we place our eye in any place in the space (x,y,z) and we select any of the viewable rays by direction (θ,Φ) as well as a band of wavelengths λ that we wish to consider (see Figure 10.2).

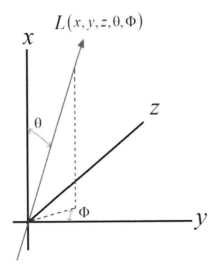

Figure 10.1   Parameterization of one ray in 3D by position (x, y, z) and direction (θ,Φ).

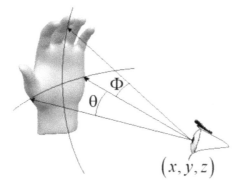

Figure 10.2   Plenoptic function describes all the images visible from a particular position.

The plenoptic function describes all of the scene information visible from a particular viewpoint. In other words, plenoptic function represents all light moving through the scene and is a function of a higher dimensionality than is required for image-based rendering.

In Levoy and Hanrahan (1996), a reduction of the dimensionality to a 4D subset was proposed, making the plenoptic function more manageable for processing of digital images. Time was discarded for static scenes and frequency replaced with the typical three color channels.

In general, there are five dimensions (three spatial and two directional dimensions); however, Levoy and Hanrahan (1996) thought that an important insight allows removal of one of the spatial dimensions (light rays do not change in value along their direction of propagation as long as the rays do not pass through an attenuating medium). This realization allows light rays to be parameterized in terms of four dimensions instead of five. Then, camera position (x,y,z) can be replaced by only one parameter k in a one-dimensional array of monocular cameras. This allows $P(x,y,z,\theta,\Phi,\lambda,t)$ to be represented as $P(k,u,v,t)$ where (u,v) are the position on each camera plane equivalent to the ray orientation (θ,Φ) (Pereira & da Silva, 2016).

The array of monocular cameras is one of the methods for measuring light fields. The array of images measured by such a device maps to a 4D light field. Also, a second type of camera employs an array of lenslets within the optical path of a monocular camera. The light is split across different pixels based on its direction of arrival. The main lens focuses the scene on the lenslet array, and the lenslet array focuses the pixels at infinity or, equivalently, on the main lens. This measures a light field that is less intuitive than in the case of a camera array (Ng *et al.*, 2005).

10.2 JPEG Pleno Parts

JPEG Pleno is a multi-part specification including the following five parts (JPEG, Overview of JPEG Pleno, 2019):

Part 1.   Framework: Specifies the JPEG Pleno framework and the inter-relationships among the different components of the standard, i.e., representation of light field, point-cloud and holographic modalities, and system-related aspects.

Part 2.   Light field coding: Specifies the coding technology for light field modalities and metadata.

Part 3.   Conformance testing: Defines conformance testing for the standardized technologies covered by the JPEG Pleno framework. In other words, this part specifies the processes to check the conformance of implementations of normative parts of JPEG Pleno.

Part 4.   Reference software: Provides non-optimized reference implementations for the standardized technologies within the JPEG Pleno framework for purpose of reference for prospective implementers of the standard and compliance testing.

Following the designed framework, it is expected that other parts will follow, notably:

Part X.   Point cloud coding: Should specify point cloud coding technologies and associated file format elements and metadata; in July 2018, the identification of use cases and definition of associated requirements was still ongoing (ISO/IEC-JTC1/SC29/WG1, 2018).

Part Y.   Hologram coding: Should specify holographic data coding technologies and associated file format elements and metadata; in July 2018, the identification of use cases was still ongoing (ISO/IEC--JTC1/SC29/WG1, 2018).

Part Z.   Quality assessment: Should specify quality assessment protocols as required by the novel imaging modalities under consideration, which require quality assessment paradigms going beyond those currently available.

10.3 Holographic Imaging

In holography, the directions of light from a 3D scene are reconstructed to produce a 2D record of a 3D scene with the strongest perceptual cue to be

Figure 10.3   Holographic data (Xing, 2015).

interpreted by the human brain as depth. Holograms deflect and focus light as prisms, lenses, and mirrors do (for one color at a time), which diffracts light into an image. A holographic image can be seen by looking into an illuminated holographic print or by shining a laser through a hologram and projecting the image onto a screen. Figure 10.3 shows an example of holographic data (Xing, 2015).

Current solutions for 3D displays like those based on optical light fields (Levoy & Hanrahan, 1996), (Yamaguchi, 2016) can only provide a subset of the required visual cues due to inherent limitations, such as a limited amount of discrete views. By contrast, digital holography (Reichelt, et al., 2010) is considered to be the ultimate display technology since it can account for all visual cues, including stereopsis, occlusion, non-Lambertian shading, and continuous parallax.

Representing the full light wavefront avoids eye vergence – accommodation conflicts (Hoffman, Girshick, Akeley, & Banks, 2008). Much of prior research efforts on holographic display systems focused on the hardware challenges, namely designing and developing the relevant optical and electronic technologies (Yaraş, Kang, & Onural, 2010), (Bider *et al.*, 2019), (JPEG, JPEG Pleno Holography, 2018).

An aspect that has received much less attention by comparison to hardware designs is how the potentially huge data volumes should be handled for efficient generation, storage, and transmission. The signal statistics of a hologram differ considerably from regular natural image and video. Hence, conventional representation and encoding algorithms, such as the standard JPEG and MPEG families of codecs, are suboptimal.

JPEG Pleno targets a standard framework for the representation and exchange of new imaging modalities such as light-field, point-cloud, and holographic imaging.

10.4 Point Cloud Imaging

A point cloud is a collection of data points which is defined by a given coordinate systems. Each point has 3D (x, y, and z) data that represents real-world objects information of every part of the object being measured. The set of points will define a three-dimensional model of an object which is represented by X, Y, and Z coordinates. These points are identified by its position in space and its color characteristics.

Point cloud images are generally produced by 3D scanners. Points in a point cloud are always located on the external surfaces of visible object because these are the spots where ray of light from the scanner is reflected from an object (Zhang, 2013).

10.5 Light-Field Imaging

Light fields are typically produced by capturing a scene either via an array of cameras or by a single and more compact sensor augmented by microlenses, to sample individual rays of light emanating from different directions. Light-field imaging strives to capture all the information of each light ray in a volume (wavelengths, polarization, intensity, and direction). With light-field photography, an image can be focused after it is taken (JPEG, Overview of JPEG Pleno, 2019), (JPEG, JPEG Pleno Light Field, 2018).

Lytro launched the world's first consumer light-field camera, which shoots pictures that can be focused long after they are captured, either on the camera itself or online. Light-field cameras also demand serious computing power, challenge existing assumptions about resolution and image quality, and are forcing manufacturers to rethink standards and usability. Perhaps most important, these cameras require a fundamental shift in the way people think about the creative act of taking a photo (Raytrix, 2019).

A light-field camera aims to measure the intensity and direction of every incoming ray. With that information, you can generate not just one but *every possible image* of whatever is within the camera's field of view at that moment. For example, a portrait photographer often adjusts the lens of the camera so that the subject's face is in focus, leaving what is behind purposefully blurry. Others might want to blur the face and make a tree in the background razor sharp. With light-field photography, you can attain either effect from the very same snapshot.

The Lytro camera uses a thin sheet containing thousands of microlenses, which are positioned between a main zoom lens and a standard 11-megapixel digital image sensor. The main lens focuses the subject onto the sheet of

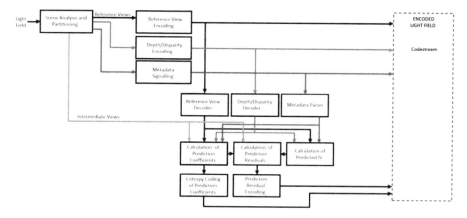

Figure 10.4 JPEG Pleno light fields codec architecture (Ebrahimi, 2019).

microlenses. Each microlens, in turn, focuses on the main lens, which, from the perspective of a microlens, is at optical infinity.

Light rays arriving at the main lens from different angles are focused onto different microlenses. Each microlens, in turn, projects a tiny blurred image onto the sensor behind it. The light contributing to different parts of those tiny blurred images comes from rays that pass through different parts of the main lens. In this way, the sensor records both the position of each ray as it passes through the main lens (the x and y) and its angle (θ and Φ) in a single exposure.

Figure 10.4 shows the JPEG Pleno light field codec architecture (Ebrahimi, 2019).

10.6 JPEG Pleno Test Conditions

JPEG Pleno (JPEG Pleno, 2019) shall support high dynamic ranges, wide color gamuts, XYZ color space, ICC profiles, transparency, and opacity. This standard shall provide the highest possible compression efficiency when compared to the raw representation and considering the available state-of-the-art solutions in the literature including the human visual system. Also, the standard should have some degree of backward compatibility with previous deployed formats (JPEG, JPEG 2000, etc.) (ISO/IEC-JTC-1/SC-29/WG1, 2015).

The standard should provide the means to trade off compression efficiency with other functionalities/features, notably random access, complexity, and scalability in terms of content, quality (SNR), spatial, depth, temporal, and spectral resolution – also in number of viewing angles, viewing angle range, and complexity.

The JPEG Pleno standard should provide the tools to achieve error resilience, both in terms of bit errors and packet losses, for a large set of networks and storage devices. Error resilience should consider graceful degradation and graceful recovery, associated with all or only parts of the content. In addition, it shall provide the means for editing and manipulation such as cutting and pasting parts in a sequence (either all or just those which are chosen). The standard should allow for low complexity in encoding and decoding, while simultaneously enabling low end-to-end content processing complexity (postproduction, rendering, etc.). Complexity needs to be considered in terms of computational complexity, memory complexity, and parallelizability.

The JPEG Pleno standard shall provide the appropriate content description tools for efficient search, retrieval, filtering, calibration, etc., of content in the various imaging modalities. Strategies and technical solutions shall fit in the JPSearch and JPEG Systems frameworks. Also, the JPEG Pleno standard shall provide the means to guarantee the privacy and security needs associated with content in the various imaging modalities. Strategies and technical solutions shall fit in the JPEG Privacy & Security and JPEG Systems frameworks.

The file format shall be consistent with the JPEG Systems specifications. It should enable additional tools to operate on JPEG Systems conforming files. The file format shall allow future extensions within the JPEG Systems framework. The standard should support distributed processing and rendering functionality as well as low latency and real-time processing implementations.

10.6.1 Holographic Data

Holograms can be generated computationally using the principles of light wave propagation. Also, they can be captured as actual physical measurements obtained typically by modulations of amplitude and phase. Holograms can be classified by the use case into holograms for visualization and microscopy and interferometry holograms (ISO/IEC JTC 1/SC 29/WG1 N88015, July 2020). The main use cases of JPEG Pleno holography are as follows.

1. Holographic microscopy (Kim, 2011) supports a large depth of field and it enables the visualization of transmissive objects. It records the light wavefront originating from an object to reconstruct the hologram using a numerical algorithm instead of acquiring the projected image of the object recorded in common microscopy. Examples of life science

applications also include monitoring the viability of cell cultures in suspensions, automating multi-well plate screening devices to measure cell density and cell coverage of adherent cell cultures, and supporting simultaneous fluorescent and holographic cell imaging.

2. Holographic tomography makes a tomogram with the refractive index (RI) and the three-dimensional location which can be obtained by calculating phase shift in hologram taken around the specimen 360°. When light passes through the object, the diffraction happens according to the object's own RI, and some properties of light including wavelength, phase shift, etc., also change. If the light that passes through the object is mixed up with the original light (reference), we can observe brightness changes in images according to the change of phase shift. This is a fundamental theory of phase contrast microscopy, and holographic tomography is an extension of holographic microscopy.

3. Holographic interferometry (Vest, 1979) allows for a quantitative comparison of two states of an arbitrary scattering, reflective, or transmissive object subject to some change. It visually reveals temporal changes (e.g., deformations, displacements, and modifications of the refractive index) without damage. The underlying principle is that incident light is reflected by the material at different angles before and after the change under consideration. Thus, holographic interferometry is widely used for non-destructive testing. It is also used in special digital cameras such as those in space and nuclear power station related applications, deep ocean exploration, and holographic endoscopes.

4. Holographic displays can realize autostereoscopic rendering without vergence-accommodation conflict (VAC) (Huffman D. C., 2008); this is because all the 3D depth cues perceived by humans in the real world are embedded in the holographic signal. The holographic display can be implemented in a variety of ways, including holographic TVs, table-top holographic displays, holographic projection systems, and holographic head-mounted displays (HMDs). The quality of holographic displays is associated with the so-called space-bandwidth product (SBP), which is a measure of the data capacity of electro-optical devices such as spatial light modulators (SLMs) (Lohmann *et al.*, 1996), (Goodman, 2004). In this context, overcoming the SBP constraints and limitations is regarded as a critical factor to realize a practical holographic display to reconstruct objects with both reasonable size and field of view (FOV). The pixel pitch of the top panels currently used in current TVs is close to 100 μm. For holographic displays, a pixel pitch of approximately 1 μm is required to provide a viewing angle of approximately 30° (Blinder

et al., 2019). Therefore, a holographic display with the same size of a current TV would require approximately 10,000 (100 × 100) times more data, which means highly effective compression mechanisms are required. Consequently, it is to be expected that the first products to be released will have humble SBPs. Hence, early products are expected to be the first in HMD and automotive windshield project system markets.

5. Holographic printing simultaneously offers the texture quality and spatial resolution of existing pictures and the 3D characteristics of holograms (Yamaguchi, Ohyama, & Honda, Holographic three-dimensional printer: new, 1992). The latter enables depth cues and parallax to be provided, unlike in pictures. Holographic printing uses a laser to record captured discrete viewpoint images (holographic stereogram printing) or wavefront (wavefront printing) in holographic material. The holographic stereogram can be best represented and compressed as a light field (JPEG Pleno, 2019). Holographic optical elements (HOEs) are used to perform the same functions as lenses, mirrors, gratings, diffusers, etc., and are a good example of holographic printing; they can also combine several functions together, which is not possible with conventional optical elements. Hence, they hold the promise for extreme miniaturization of certain complex optical or digital image processing steps.

Holographic specific requirements:
The JPEG Pleno shall support sufficiently high space-bandwidth products for capture and displays systems to meet image size and viewing angle requirements. In addition to the generic color representation requirements, the standard shall support color specifications typically needed for holographic rendering systems. Besides, it shall incorporate signaling syntax to describe capturing and/or rendering conditions in order to transform content from capturing reference to rendering reference in the processing pipeline and support signaling of information necessary for display specific processing steps.

Below is a list of requirements that emerge from the use cases. The standard shall support:

1. amplitude modulation hologram, phase modulation hologram, and complex modulation holograms;
2. one or more of the amplitude-only, phase-only, amplitude-phase, and real-imaginary formats;

3. to support large FOV and a large viewing angle produced by the large space-bandwidth product holograms, the standard shall be able to support large spatial resolution of the hologram;
4. holographic data with bit depths ranging from bi-level up to 16-bit integer precision;
5. at least 32-bit IEEE 754 floating-point representation format;
6. multiple components, including RGB;
7. hologram sequences both in time and space;
8. high compression efficiency considering the state-of-the-art;
9. quality control, notably for machine vision analysis in the context of interferometry and holographic microscopy/tomography;
10. near-lossless and lossy coding up to high quality coding;
11. coding types that should be supported:
    a. lossless coding;
    b. perceptual lossless coding (the reconstructed holograms are visually indistinguishable from the original reconstructed holograms);
12. allow decoding of subsets of the complete codestream to enable selecting an aperture window in space-frequency domain of the hologram;
13. resilient against bit errors and packet losses;
14. facilitate efficient implementation on GPU and multi-core architectures;
15. facilitate efficient implementation in hardware platforms;
16. be compliant with the JPEG Pleno framework architecture as defined in Part 1 (ISO/IEC 21794);
17. enable signaling relevant metadata such as capture parameters, calibration data (e.g., geometrical setup), half/full parallax, wavelengths, image sensor (CCD/CMOS) information, horizontal and vertical pixel pitch, pixel shape, microscope configuration, spectral channel information, display and rendering parameters, and information about additional processing steps;
18. provide support for JPEG Privacy and Security tools.
19. Efficient change of depth of field or viewpoint, refocusing, relighting, navigation, rotation, and enhanced analysis of objects without transcoding of content.

The standard also should support the decoding of different levels of quality (SNR), spatial resolution, depth resolution, temporal resolution, spectral resolution, and FoV from a subset of the codestream and coding tools that exploit the human visual system properties to achieve maximal visual quality for reconstructed holograms. Also, it should remain within reasonable bounds

Table 10.1 Target bit rates for the JPEG Pleno holography test set.

Dataset	Target bit rate (bpp)					
Monochrome holograms	0.1	0.25	0.5	1	2	4
Color holograms	0.3	0.75	1.5	3	6	12

with respect to near-future computing and memory capabilities and allow for staged processing to reduce the internal decoder bandwidth consumption.

Bit rates and rate metrics:
Target bit rates for the experiments are provided in Table 10.1. The bit rates for the color holograms are three times as large as that for the monochrome holograms since it is assumed that the color planes are encoded separately without exploiting potential correlations. Hence, every color plane is attributed by one-third of the bit budget during encoding with the anchor encoders.

The bit rate, specified in Table 10.4, accounts for the total number of bits necessary for generating the encoded file (or files), out of which the decoder can reconstruct a lossy or lossless version of the entire input hologram. The main rate metric is defined as the number of bits per sample (pixel) as

$$\text{Bitrate (bpp)} = \text{Total Number of Bits / Number of Samples.} \quad (10.1)$$

In this case, a sample can be complex valued. In this case, the number of bits per sample is the sum of the number of bits for the real and imaginary components.

Quality metrics:
Intuitively, the quality of a compression scheme can be characterized by how close the decompressed image is to the original. There exist metrics that apply to holographic compression data. These metrics are described as follows.

SNR and PSNR: The signal-to-noise ratio (SNR) is the ratio of the power of the signal to the power of the noise affecting the quality of the signal, while the peak signal-to-noise ratio (PSNR) is defined as the ratio between the maximum possible power of a signal and the power of noise. The SNR (in dB) is calculated on the complex valued wavefield in the hologram plane and is given by

$$\text{SNR} = 10\log_{10}\left(\frac{\sum_{i=1}^{A}\sum_{j=1}^{B}\left|X[i,j]\right|^{2}}{\sum_{i=1}^{A}\sum_{j=1}^{B}\left|X[i,j]-\hat{X}[i,j]\right|^{2}}\right) \quad (10.2)$$

where X and $\hat{X}$ are the reference and the reconstructed holograms, respectively. The PSNR is used for evaluating the quality of reconstructions at the object plane

$$\text{PSNR} = 10\log_{10}\left(\frac{AB(2^n - 1)}{\sum\limits_{i=1}^{A}\sum\limits_{j=1}^{B}\left|X[i,j] - \hat{X}[i,j]\right|^2}\right) \tag{10.3}$$

where n is the bit depth and X and $\hat{X}$ are the reference and the reconstructed holograms, respectively.

SSIM: The Structural SIMilarity (SSIM) index (Wang, Bovik, Sheikh, & et al., 2004) is a full-reference perceptual metric to quantify the visual quality degradation measured by perceived change in structural information. For complex valued data, the SSIM is obtained as the mean of the SSIM of the real and imaginary parts. The SSIM index is bounded between -1 and 1, where values closer to 1 indicate high correlation and better perceptual quality, while values closer to -1 indicate negative correlation. For compression, the range of values will lie closely in the range $0-1$. The SSIM index is calculated on various windows of an image. The measure between two windows x and y of common size $N \times N$ is

$$\text{SSIM}(x,y) = \frac{\left(2\mu_x\mu_y + c_1\right)\left(2\sigma_{xy} + c_2\right)}{\left(\mu_x^2 + \mu_y^2 + c_1\right)\left(\sigma_x^2 + \sigma_y^2 + c_2\right)} \tag{10.4}$$

where μ_x, μ_y, σ_x, and σ_y are the mean and standard deviation of x and y, respectively, and $c_1 = (k_1 L)^2$, $c_2 = (k_2 L)^2$ with $k_1 = 0.01$, $k_2 = 0.03$ by default and L is the dynamic range of the pixel value.

Bjøntegaard metric: The Bjøntegaard metric compares the rate-distortion performance of two coding solutions across some rate/distortion region by computing the surface area that lies between the rate-SNR/SNR-rate curves of the two codecs, where the rate axis is logarithmically scaled (Bjontegaard, 2001).

VIFp: The visual information fidelity in pixel domain (VIFp) (Sheikh & Bovik, 2006) is a faster implementation of the visual information fidelity (VIF) which performs multi-scale analysis in spatial domain instead of originally utilized wavelet domain in VIF. In its core, VIF approaches the overall visual process through the human visual system (HVS) as a baseline distortion channel which is added to every input data and models it using a stationary, zero mean, additive white Gaussian noise. Next, the mutual information is calculated between the source model (represented by the natural scene statistics) and the test image after adding the HVS baseline distortion. The value then is normalized by the value of another mutual information similarly calculated for the reference image. VIF is bounded below by 0, which indicates that all information about the reference image has been lost in the distortion channel. In case of no distortion (reference compared to itself), VIF is exactly unity. However, its upper bound is not limited to 1. For example, in case of a linear contrast enhancement of the reference image that does not add noise to it will result in a VIF value larger than one.

SNR of first-order wavefield: For off-axis holograms, the relevant information is encoded in the first-order wavefield. The fidelity of the compressed first-order wavefield is measured by the signal-to-noise ratio (SNR) metric as

$$\mathrm{SNR} = 10\log_{10}\left(\frac{\displaystyle\sum_{u=-B_u}^{B_u}\sum_{v=-B_v}^{B_v}\left|U_f[u,v]\right|^2}{\displaystyle\sum_{u=-B_u}^{B_u}\sum_{v=-B_v}^{B_v}\left|U_f[u,v]-\hat{U}_f[u,v]\right|^2}\right) \tag{10.5}$$

where the demodulated first-order wavefield in the frequency domain and its compressed version are denoted by U_f and $\hat{U}_f$, respectively. The bandwidths of the first-order term are represented by the sets $[-B_u, B_u]$ and $[-B_v, B_v]$.

RMSE of retrieved phase: The root mean squared error (RMSE) of the retrieved phase is calculated by

$$\mathrm{RMSE} = \sqrt{\sum_{i=L_a}^{L_b}\sum_{j=B_a}^{B_b}\frac{\left(\Phi[i,j]-\hat{\Phi}[i,j]\right)^2}{\left(L_a-L_b\right)\left(B_a-B_b\right)}} \tag{10.6}$$

where $[L_a, L_b]$ and $[B_a, B_b]$ describe the spatial boundary of the phase function Φ and $\hat{\Phi}$ retrieved from the original hologram and the compressed hologram, respectively.

Hamming distance: For binary holograms X, the average Hamming distance between the compressed hologram $\hat{X}$ is given by

$$H = \frac{1}{AB}\sum_{i=1}^{A}\sum_{j=1}^{B}\left(X[i,j]\oplus\hat{X}[i,j]\right)^2 \tag{10.7}$$

where $\oplus$ is the XOR operation.

Color information metric: Currently, no validated procedures exist to de-correlate color information in holography. For compression using anchor codecs, the three color channels are compressed independently. For quality evaluation, color holograms are not converted to another color space. The quality metrics are computed for each color channel independently and the arithmetic mean is calculated and

$$M = \frac{M_G + M_R + M_B}{3} \qquad (10.8)$$

where M_G, M_R, and M_B are the quality metrics for red, green, and blue channels, respectively.

Configuration quality metrics and anchor codecs:
Table 10.2 shows the quality metrics proposed according to the type of hologram.

Until now, no standards have been specified for coding of holographic content. The anchor codecs selected for reference purposes are H.265/HEVC (configured in intra-frame mode and HM version 16), JPEG 2000, and JBIG-1 originally designed for natural images or binary content. The anchor codecs are tested in two pipelines, one performing the encoding in the hologram in the hologram plane, and the other in the object plane. Visual quality assessment is performed in both planes, except for the subjective visual quality assessment, which is solely performed in the object plane. Metrological data quality is measured directly on the metrological data extracted from the uncompressed (original) and compressed holograms. Table 10.3 shows the reference anchor codecs and their employment during testing.

Table 10.2 Quality metrics according to the type of hologram.

| | Hologram type | | | | |
| | Higher precision | | Binary | | Metrological |
Metric	Hologram plane	Object plane	Hologram plane	Object plane	--
SNR	Yes	--	Yes	--	Yes
PSNR	--	Yes	--	Yes	--
SSIM	Yes	Yes	--	Yes	--
VIFp	--	Yes	--	Yes	--
Hamming distance	--	--	Yes	--	--
SNR of first order	--	--	--	--	Yes
RMSE of retrieved phase	--	--	--	--	Yes

Table 10.3 Anchor codecs for reference purposes.

	Hologram type			
	Higher precision		Binary	
Anchor codec	Hologram plane	Object plane	Hologram plane	Object plane
H.265/HEVC	Yes	Yes	--	--
JPEG 2000	Yes	Yes	--	--
JBIG-1	--	--	Yes	--

10.6.2 Point Cloud

A point cloud refers to a set of data items representing positions of points in space, expressed in a given coordinate system, usually three-dimensional. This geometrical data can be accompanied with per-point attributes of varying nature (e.g., color or temperature). Such datasets are usually acquired with a 3D scanner, LIDAR, or created using 3D design software and can subsequently be used to represent and render 3D surfaces. Combined with other types of data (like light-field data), point clouds open a wide range of new opportunities for immersive browsing and virtual reality applications (ISO/IEC JTC 1/SC29/WG1 N88014, 2020). Some use cases are:

1. immersive communication;
2. cultural heritage;
3. industrial;
4. holography.

Test set material:
Test set will be organized in human beings and small inanimate objects. The selected human beings' point clouds are organized as full bodies and upper bodies.

   Full bodies[1]: Two common voxelized test sequences in this dataset are longdress and soldier (d'Eon, Harrison, Myers, & et al., 2020), (JPEG, JPEG Pleno Database: 8i Voxelized Full Bodies (8iVFB v2) – A Dynamic Voxelized Point Cloud Dataset, 2017). In each sequence, the full body of a human subject is captured by 42 RGB cameras configured in 14 clusters (each cluster acting as a logical RGBD camera), at 30 fps, over a 10-second period. One spatial resolution is provided for each sequence: a cube of 1024 × 1024 × 1024 voxels; this results in a geometry precision of 10 bits. For each sequence, the cube is scaled so that it is the smallest bounding cube that contains every frame in the entire sequence. Since the height of the subject

---

[1] This dataset can be found at: https://jpeg.org/plenodb/pc/8ilabs/

Table 10.4   Full Bodies point cloud properties.

Name	Frame number	File name	Number of points	Geometry precision
longdress	1300	longdress_vox10_1300.ply	857,966	10 bits
soldier	0690	soldier_vox10_0690.ply	1,089,091	10 bits

Table 10.5   Upper Bodies point cloud properties.

Name	Frame number	File name	Number of points	Geometry precision
ricardo10	82	ricardo10_frame0082.ply	1,414,040	10 bits
phil0	139	phil9_frame0139.ply	356,258	9 bits

is typically the longest dimension, for a subject 1.8 m high, a voxel with a geometric precision of 10 bits would be approximately 1.8 m/1024 voxels ≈ 1.75 mm per voxel on a side. As these are dynamic point clouds, only a single frame specified in Table 10.4 will be used.

Upper body[2]: The dynamic voxelized point cloud sequences in this dataset are the Microsoft Voxelized Upper Bodies (MVUB) (Loop, Cai, Orts, & et al., 2016). Two subjects in the dataset, known as ricardo10 and phil9, are used for testing. The heads and torsos of these subjects are captured by four frontal RGBD cameras, at 30 fps, over a 7- to 10-second period for each. For the JPEG Pleno Point Cloud Coding test set, a single frame was selected from each sequence as indicated in Table 10.5. Two spatial resolutions are provided for each sequence: a cube of $512 \times 512 \times 512$ voxels and a cube of $1024 \times 1024 \times 1024$ voxels, indicating a geometry precision of 9 and 10 bits, respectively. As these are dynamic point clouds, only a single frame from each sequence as specified in Table 10.6 will be used.

Inanimate objects[3]. These point clouds all have 24-bit RGB color information associated with each point. The romanoillamp and bumbameuboi point clouds have been selected for the JPEG Pleno Point Cloud Coding test set to represent cultural heritage applications. Voxelized versions of these point clouds are available in the JPEG Pleno Point Cloud Coding test set. The voxelized version of the romanoillamp has been voxelized to a geometry precision of 10 bits. In addition, École Polytechnique Fédérale de Lausanne (EPFL) donated the guanyin and rhetorician point clouds also representing cultural heritage application. Both of these point clouds have been voxelized

---

[2] This dataset can be found at: https://jpeg.org/plenodb/pc/microsoft/

[3] University of San Paulo, Brazil supplied a set of point clouds to JPEG related to cultural heritage. This dataset can be found at: http://uspaulopc.di.ubi.pt/

Table 10.6   Small inanimate objects point cloud properties.

Point cloud name	Frame number	Point cloud file name	Number of points	Geometry precision
romanoillamp	--	romanoillamp_Transform_Denoise_vox10.ply	638,071	10 bits
bumbameuboi	--	bumbameuboi_Rescale_Denoise_vox10.ply	113,160	10 bits
guanyin	--	guanyin_vox10.ply	2,286,975	10 bits
rhetorician	--	rhetorician_vox10.ply	1,732,600	10 bits

to a geometry precision of 10 bits. The properties of these point clouds are described in Table 10.6.

Subjective quality evaluation:
For subjective evaluation, point clouds will be rendered and displayed as follows:
Rendering:
- Point clouds will be rendered to a video using CloudCompare software [CloudCompare19] with the following settings:
  ○ Default parameters except for point size.
  ○ Point size needs to be determined for each sequence, by experimentation to render as far as possible the perception of a watertight surface.
- Point clouds will be rendered against a black background without surface reconstruction or, in other words, as individual points.
- Point clouds should be rendered for, and viewed on, 2D monitors with resolution of at least 1920 × 1080 and ideally of resolution 3840 × 2160 and a color gamut of sRGB or wider. Point clouds should be rendered such that the intrinsic resolution of the point cloud matches the display resolution. This requires as far as possible that point density of the rendered point cloud be such that no more than one point should occupy a single pixel of the displayed video. This may require the display of cropped sections of the point cloud with suitably adjusted view paths.
  ○ Customization for resolution of the frames is by adjusting the edge of 3D view window to the desired resolution.
  ○ In order to get full resolution of 2D monitors, press F11 (full screen 3D view) before choosing view and rendering.
Video production:
- To create the video, the camera will be rotated around the object to create a set of video frames that allow for viewing of the entire object. The exact path used for the video sequence will vary depending on the

content type and is only known to the Chair and Co-chair of the Ad Hoc Group on JPEG Pleno – Point Cloud, Stuart Perry (University of Technology Sydney, Australia) and Luis Cruz (University of Coimbra, Portugal). The view paths are not revealed to proponents to avoid proposals being explicitly tailored according to content view paths.

- Videos will be created such that each video is of 12-second duration at a rate of at least 30 fps. As far as possible, the video view paths should be adjusted such that the duration of the video is approximately 12 seconds, while the degree of apparent motion between frames is the same for all stimuli videos.
- The video frames will be then visually losslessly compressed (i.e., constant rate factor equal to 17) with the HEVC encoder (using FFmpeg software), producing an animated video of 30 fps with a total duration of 12 seconds.

Viewing conditions:
- Viewing conditions should follow ITU-R Recommendation BT.500.13 (ITU-R-BT.500-13, 2012). MPV video player will be used for displaying the videos. Displays used in the subjective testing should have anti-aliasing disabled.

Considerations for subjective evaluation:
Prior to subjective evaluation, there will be a training period to acclimatize participants to the task to be performed. This training period involves showing participants video sequences similar to the ones used in the test but not the point clouds used in the subsequent test sequences. The phil9 point cloud will be used for this purpose. Participants are requested to score the perceived quality of the rendered point cloud in relation to the uncompressed.

The DSIS simultaneous test method will be used with a five-level impairment scale, including a hidden reference for sanity checking. Both the reference and the degraded stimuli will be simultaneously shown to the observer, side-by-side, and every subject asked to rate the visual quality of the processed with respect to the reference stimulus. To avoid bias, in half observer presentations, the reference will be placed on the right and the degraded content on the left side of the screen and vice-versa for the rest of the evaluations.

Outlier detection algorithm based on ITU-R Recommendation BT.500-13 (ITU-R-BT.500-13, 2012) should be applied to the collected scores, and the ratings of the identified outliers were discarded. The scores are then averaged to compute mean opinion scores (MOS) and 95% confidence intervals (CIs) computed by assuming a Student's *t*-distribution.

Rate and rate-distortion evaluation:
The main rate metric is the number of bits per point (bpp) defined as

$$\text{Bits per point} = \text{NTotBits/NTotPoints} \qquad (10.9)$$

where NTotBits is the number of bits for the compressed representation of the point cloud and NTotPoints is the number of points in the input point cloud. The bits per point used to code geometry and the bits per point used to code attributes should also be reported.

Scalability:
Scalability capabilities shall be assessed by encoding the point cloud using scalable and non-scalable codecs at a variety of rate points. The relative compression efficiency will be assessed using Bjontegaard delta metrics (rate and PSNR) (Bjontegaard, 2001) between scalable and non-scalable codecs. The JPEG Committee will compute rate–distortion (RD) curves using the D1, D2, and point-to-point attribute measure metrics described below to allow different scalability requirements to be tested. Scalable codecs will have rate and distortion values obtained when decoding at each level of scalability. If a scalable codec has a non-scalable mode, then rate and distortion values will also be obtained for the codec in the non-scalable mode. Reports on scalability should report on bits required to achieve each specific scalability level.

Random access:
The ROI random access measure shall be measured, for a given quality, by the maximum number of bits required to access a specific region of interest (ROI) of the point cloud divided by the total number of coded bits for the point cloud as

$$\text{ROI Random Access} = \frac{\text{Total amount of bits that have to be decoded to access an ROI}}{\text{Total amount of encoded bits to decode the full point cloud}}.$$

$$(10.10)$$

A set of predetermined ROIs will be defined by the JPEG Committee. A codec must provide the entire ROI but may provide more points. The result should be reported for the specific ROI type that gives the worst possible result for the ROI random access.

Quality metrics:
Quality metrics measure the similarity of the decoded point cloud to a reference point cloud using an objective measure. The quality metrics considered for coding of point cloud images are as follows.

Point-to-geometry D1 metric[4]: The point-to-point geometry metric (D1) is based on the geometric distance of associated points between the reference point cloud and the content under evaluation. In particular, for every point of the content under evaluation, a point that belongs to the reference point cloud is identified, through the nearest neighbor algorithm. Then, an individual error is computed based on the Euclidean distance. This error value is associated with every point, indicating the displacement of the distorted point from the reference position. The error values for each point are then summed to create a final measure. In accordance with the description given in (ISO/IEC JTC1/SC29/WG11 MPEG2016/n18665, 2019), the point-to-point error measure D1 is computed as follows.

First, determine an error vector $E(i, j)$ denoting the difference vector between the identified point a_j in reference point cloud A to the corresponding point b_j (identified by the nearest neighbor algorithm) in point cloud B. The length of the error vector is the point-to-point error, i.e.,

$$e_{B,A}^{D1}(i) = \left\| E(i,j) \right\|_2^2 \tag{10.11}$$

The point-to-point error ($D1$) for the whole point cloud is then defined as the arithmetic mean $e_{B,A}^{D1}$ of the errors $e_{B,A}^{D1}(i)$. This error is expressed as the symmetric PSNR using the following formula:

$$\text{PSNR} = 10 \frac{3 * \text{peak}^2}{\max\left(e_{B,A}^{D1}, e_{A,B}^{D1}(i)\right)} \tag{10.12}$$

where peak is the resolution of the model (i.e., if voxel bit depth = 10, it is 1024) and $e_{B,A}^{D1}$ is the point-to-point error when the roles of A and B are swapped in Equation (10.11). The denominator of Equation (10.12) ensures that the resultant PSNR is symmetric and invariant to which of the point clouds compared is considered the reference.

For near-lossless geometry encoding, the maximum point-to-point error, denoted here as $D1_{max}$, should be considered instead.

---

[4] The D1 metric will be computed using the software supplied by WG11 at: http://mpegx. int-evry.fr/software/MPEG/PCC/mpeg-pcc-dmetric/tree/master

Point-to-plane geometry D2 metric[5]: The point-to-plane geometry metric (D2) is based on the geometric distance of associated points between the content under evaluation, B, and the local planes fitted to the reference point cloud, A. In particular, for every point of the content under evaluation, b_j, a point that belongs to the reference point cloud, a_j, is identified through the nearest neighbor algorithm and a plane is fitted to the region centered on the identified point on the reference cloud. The normal of this plane is denoted as N_j and is computed using quadric fitting in CloudCompare (2019), including points within a radius of 5. Where a point cloud has pre-computed available information for the normal at a_j, this information may be used in place of N_j. The distance vector $E(i, j)$ is computed and projected via the dot product $\langle , \rangle$ onto the normal vector, N_j to compute the point-to-plane error (ISO/IEC JTC1/SC29/WG11 MPEG2016/n18665, 2019)

$$e_{B,A}^{D2}(i) = \left\langle E(i,j), N_j \right\rangle^2 . \tag{10.13}$$

The point-to-plane error (D2) for the whole point cloud is then defined as the arithmetic mean $e_{B,A}^{D1}$ of the errors $e_{B,A}^{D1}(i)$. This error is expressed as the symmetric PSNR using Equation (10.12).

For near-lossless geometry encoding, the maximum point-to-point error, denoted here as $D2_{max}$, should be considered instead.

Point-to-plane angular similarity metric[6]: This metric is based on the angular similarity (Alexiou & Ebrahimi, 2018) of tangent planes that correspond to associated points between the reference and the content under evaluation. In particular, for each point that belongs to the content under evaluation, a point from the reference point cloud is identified, using the nearest neighbor algorithm. Then, using the normal vectors for the reference point cloud and the point cloud under consideration, the angular similarity of tangent planes is computed based on the angle θ, which denotes the minimum out of the two angles that are formed by the intersecting tangent planes (ISO/IEC JTC1/SC29/WG 1 M81049, 2018) according to

$$\text{Angular similarity} = 1 - \frac{\theta}{\pi} \tag{10.14}$$

---

[5] The *D2* metric will be computed using the software supplied by WG11 at: http://mpegx.int-evry.fr/software/MPEG/PCC/mpeg-pcc-dmetric/tree/master

[6] The implementation used for the angular similarity metric will be that given in: https://github.com/mmspg/point-cloud-angular-similarity-metric

This error value is associated with every point, providing a coarse approximation of the dissimilarity between the underlying local surfaces.

Point-to-point attribute measure[7]: The point-to-point attribute metric is based on the error of attribute values of associated points between the reference point cloud and the content under evaluation. In particular, for every point of the content under evaluation, a point that belongs to the reference point cloud is identified, through the nearest neighbor algorithm. Then, an individual error is computed based on the Euclidean distance. For color attributes, the MSE for each of the three color components is calculated. For near-lossless coding, the maximum error should be considered instead of MSE. A conversion from RGB space to YCbCr space is conducted using (ITU-R-BT.709-6, 2015). PSNR value is then computed for each channel as

$$PSNR = 10\frac{p^2}{MSE}. \tag{10.15}$$

The maximum distortion between the two passes is selected as the final distortion. Since the color attributes for all test data have a bit depth of 8 bits per point, the peak value p for PSNR calculation is 255. PSNR values for the Y, Cb, and Cr channels are combined as

$$PSNR_{Color} = \frac{4PSNR_Y + PSNR_{Cb} + PSNR_{Cr}}{6}. \tag{10.16}$$

Target bit rates: Each test point cloud considered for subjective evaluations will have to be encoded at the target bit rates (within a tolerance of +−10%) of 0.1, 0.35, 1.0, 2.0, and 4.0 bits per point.

Anchor generation:
Anchor codecs: The anchors will be generated by the geometry-based point cloud compression[8] (G-PCC) TMC13 (ISO/IET JTC1/SC29/WG11 & WG11N18189, 2019), and video point cloud coding[9] (V-PCC) TMC2 codecs (ISO/IEC JTC1/SC29/WG11 & MPEG2019/N18190, 2019), (ISO/IET JTC1/SC29/WG11 & WG11N18475, "V-PCC Test Model v6," 2019).

---

[7] The point-to-point attribute measure metric will be computed using the software supplied by WG11 at: http://mpegx.int-evry.fr/software/MPEG/PCC/mpeg-pcc-dmetric/tree/master

[8] The software for G-PCC is available from the MPEG GitLab: http://mpegx.int-evry.fr/software/MPEG/PCC/TM/mpeg-pcc-tmc13.git.

[9] The software implementation of V-PCC to be used is available from the MPEG GitLab: http://mpegx.int-evry.fr/software/MPEG/PCC/TM/mpeg-pcc-tmc2.git

G-PCC expects geometry to be expressed with integer precision; so all content must be quantized or voxelized prior to encoding. This should be achieved by oct-tree voxelization to a depth that preserves the number of points in the original point cloud.

10.6.3 Light Field

JPEG Pleno test set is organized according to the acquisition/creation technology as this factor determines several content characteristics, notably the views baseline and amount of interview redundancy (ISO/IECJTC1/SC29/WG1/N84025, 2019).

Test set material:
A quantitative descriptor, the so-called Geometric Space-View Redundancy (GSVR) descriptor, has been used to characterize each light-field image. This section describes the test material selected for JPEG Pleno core and exploration experiments.

Lenslet Lytro Illum Camera:

The light-field images were captured using a Lytro Illum B01 (10-bit) light-field camera (Řeřábek & Ebrahimi, 2016), (Řeřábek, Yuan, Authier, & et al., July, 2015). JPEG Pleno test set includes four lenslet light-field images in the Lenslet Lytro Illum Camera dataset, notably Bikes, Danger de Mort, Stone Pillars Outside, and Fountain&Vincent

Fraunhofer High Density Camera Array (HDCA):

JPEG Pleno test set includes only one light-field image from the four initially provided in the context of the Fraunhofer HDCA (Fraunhofer HHI, March, 2017) set.

Poznan HDCA:

JPEG Pleno test set includes only one light-field image from the two provided in the context of the Poznan HDCA set, named Laboratory 1. The sequences remain the property of Poznan University of Technology, but they are licensed for free use within ISO/IEC JTC1/SC29/WG11 (MPEG) and ISO/IEC JTC1/SC29/WG01 (JPEG) for the purposes of research and development of standards (Domański, Klimaszewski, Kurc, & et al., 2014).

Stanford HDCA:

JPEG Pleno test set includes only one light-field image in the Stanford HDCA dataset from the many images available. The selected light-field image has been named Tarot Cards; it corresponds to the Tarot Cards & Crystal Ball (small angular extent) image. This dataset has positive and negative disparities. The light fields may be used for academic and research purposes but are not to be used for commercial purposes, nor should they appear in a product for sale without permission. If you use these models in a publication, please credit the Stanford Computer Graphics Laboratory (Vaish, 2008).

Synthetic HCI HDCA:

Two light-field images from the Synthetic HCI HDCA dataset are included in the JPEG Pleno sets, notably Greek and Sideboard. These light-field images were synthetically created and are made available by the Heidelberg Collaboratory for Image Processing (HCI) (HCI 4D Light Field Dataset, 2021).

Table 10.7 is a summary of the selected JPEG Pleno light-field images (ISO/IECJTC1/SC29/WG1/N84025, 2019).

Quality metrics:

RGB to YCbCr conversion: The 10-bit RGB images are converted to double precision and these double precision RGB images are converted to double precision Y, Cb, Cr using the ITU-R BT.709-6 recommendation (ITU-R-BT.709-6, 2015).

Table 10.7 Summary of the selected JPEG Pleno light-field images.

Source	Type	Light-field image name	Number of views	Spatial resolution (pixels)	Link
EFPL	LL	Bikes, Danger de Mort, Stone Pillars Outside, Fountain & Vincent 2	15 × 15	625 × 434	https://jpeg.org/ jpegpleno/p lenodb. html
Fraunhofer IIS	HDCA	Set 2 2K sub	33 × 11	1920 × 1080	https://jpeg.org/ jpegpleno/p lenodb. html
Poznan University of Technology	HDCA	Laboratory1	31 × 31	1936 × 1288	https://jpeg.org/ jpegpleno/p lenodb. html
New Stanford LF Archive	HDCA	Tarot Cards	17 × 17	1024 × 1024	https://jpeg.org/ jpegpleno/p lenodb. html
HCI	HDCA	Greek, Sideboard	9 × 9	512 × 512	https://jpeg.org/ jpegpleno/p lenodb. html

PSNR: The PSNR between the original X component and the reconstructed component $\hat{X}$ is computed as

$$PSNR = 10\log_{10}\left(\frac{2^n - 1}{\frac{1}{MN}\sum_{i=0}^{M-1}\sum_{j=0}^{N-1}\left(X[i,j] - \hat{X}[i,j]\right)^2}\right) \quad (10.17)$$

where n is the bit depth, and $M \times N$ is the component size. PSNR of the YCbCr images is computed as

$$PSNR_{YCbCr} = \frac{6 * PSNR_Y + PSNR_{Cb} + PSNR_{Cr}}{6}. \quad (10.18)$$

SSIM: The SSIM (Wang, Bovik, Sheikh, & et al., 2004) for the whole light field is computed by averaging the SSIM values of the individual views or sub-aperture images.

Bjøntegaard metric: Include the Bjøntegaard metric that provides numerical averages between RD-curves as part of the presentation of results (Bjontegaard, 2001).

Rate metrics:
The main rate metric is the number of bits per pixel (bpp) computed as

$$BPP = \frac{\text{Number of bits for the compressed representation}}{\text{Number of pixels in the whole light field}}. \quad (10.19)$$

For encoding the HDCA Set2 2K subdataset that has 33×11 views, each with 1920×1080 resolution, the total number of pixels is $33 \times 11 \times 1920 \times 1080 = 752,716,800$ pixels.

Random access metric:
The random access penalty metric is defined as the ratio of amount of encoded bits required to access an ROI and the total amount of encoded bits as

$$\text{Random Access Penalty} = \frac{\begin{array}{c}\text{Total amount of encoded bits} \\ \text{required to access an ROI}\end{array}}{\begin{array}{c}\text{Total amount of encoded bits} \\ \text{required to decode the full light field}\end{array}}.$$

$$(10.20)$$

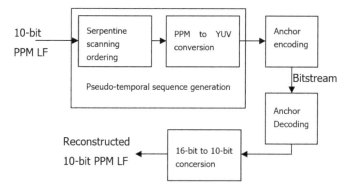

Figure 10.5   Encoding/decoding pipeline (ISO/IECJTC1/SC29/WG1/N84025, 2019).

The result should be reported for the specific ROI type that gives the worst possible result for the random access penalty metric.

HEVC anchor generation:
The selected anchor is the HEVC standard[10]. The encoding/decoding pipeline is shown in Figure 10.5.

---

[10] The x.265 implementation (x265.org) of HEVC, source version 2.3 (available from https://bitbucket.org/multicoreware/x265) should be used.

11

JPEG AIC

The purpose of the advanced image coding and evaluations (JPEG, Overview of JPEG AIC, 2019) work is to locate and evaluate new scientific developments and advancements in image coding research. Relevant topics include new compression methodologies or quality evaluation methodologies and procedures.

Currently, the work of JPEG AIC resulted in two standards (ISO/IEC 29170-1 and ISO/IEC 29170-2). ISO/IEC 29170-1 provides a framework that allows evaluating the suitability of image compression algorithms in various application domains. To this end, this standard aims to provide a set of evaluation tools that allow multiple features of such codecs to be tested and compared. The second standard, ISO/IEC 29170-2, defines a set of evaluation procedures, similar to how the ITU-R BT.500-13 (ITU-R-BT.500-13, 2012) standard did for the subjective quality evaluation with television standards (JPEG, Overview of JPEG AIC, 2019), (JPEG, Workplan & Specs of JPEG AIC, 2018).

JPEG provides very good quality of reconstructed images at low or medium compression (JPEG, Documentation on JPEG AIC, 2018) (i.e., high or medium bit rates, respectively), but it suffers from blocking artifacts at high compression (low bit rates). Erick Van Bilsen has developed an experimental still image compression system known as advanced image coding (AIC) that encodes color images and performs much better than JPEG and close to JPEG-2000 (Van Bilsen, 2018). AIC combines intra-frame block prediction from H.264 with a JPEG-style discrete cosine transform, followed by context adaptive binary arithmetic coding (CABAC) used in H.264. The aim of AIC is to provide better image quality with reduced complexity. It is also faster than existing JPEG 2000 codecs (Veerla, Zhang, & Rao, 2012).

11.1 Advanced Image Coding

Advanced image coding (AIC) is an experimental still image compression system that combines algorithms from the H.264 and JPEG standards. More specifically, it combines intra-frame block prediction from H.264 with a JPEG-style discrete cosine transform, followed by context adaptive binary arithmetic coding as used in H.264. The result is a compression scheme that performs much better than JPEG and close to JPEG-2000.

11.2 AIC Characteristics

1. For photographic images, AIC performs much better than JPEG and close to JPEG-2000. For typical bit rates, AIC sometimes even outperforms JPEG-2000.
2. For graphic images, the gap between JPEG-2000 and AIC grows, in the favor of JPEG-2000.
3. For small images, the gap between JPEG-2000 and AIC also grows; however, this time in the favor of AIC. For image sizes up to 100×100 (or 10,000 pixels), AIC performs much better for typical bit rates. This would make an AIC-like codec ideal for images on web pages.
4. YCbCr format is 4:4:4.
5. Uses the same color conversion method as in JPEG reference software.
6. No subsampling – higher quality/compression ratios.
7. Nine prediction modes as in H.264 video codec.
8. Blocks are predicted from previously decoded blocks.
9. The codec uses DCT to transform 8 x 8 residual blocks instead of transform coefficients as in JPEG format (JPEG, Overview of JPEG, 2018).
10. Uniform quantization is used in the transform coefficients.
11. Floating-point algorithm is used.
12. The coefficients are encoded in scan-line order.
13. Uses CABAC as in H.265 with several contexts.
14. AIC is somewhat slower than JPEG but faster than JPEG-2000, even without speed optimizations in the software.

10.3 AIC Codec

Figure 11.1 shows the AIC codec. Both encoder and decoder share the same functional elements (Van Bilsen, 2018).

The input RGB image to the encoder is converted into the channels YCbCr. This operation makes the chrominance channels more suitable to be coded because, in this color space, chrominance has less entropy. Equations (11.1) and (11.2) show the color conversion matrices.

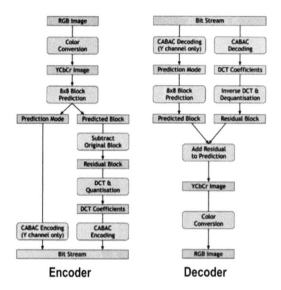

Figure 11.1 AIC codec (Van Bilsen, 2018).

$$\begin{bmatrix} Y \\ Cb \\ Cr \end{bmatrix} = \begin{bmatrix} 0.299 & 0.587 & 0.114 \\ -0.169 & -0.331 & 0.500 \\ 0.500 & -0.419 & -0.081 \end{bmatrix} \begin{bmatrix} R \\ G \\ B \end{bmatrix} \qquad (11.1)$$

$$\begin{bmatrix} R \\ G \\ B \end{bmatrix} = \begin{bmatrix} 1 & 0 & 1.402 \\ 1 & -0.344 & -0.714 \\ 1 & 0.772 & 0 \end{bmatrix} \begin{bmatrix} Y \\ Cb \\ Cr \end{bmatrix}. \qquad (11.2)$$

Every channel is divided into blocks of 8 × 8 pixels. Then, each block is predicted from previously encoded and decoded blocks in their correspondent channel. Nine prediction modes are used to predict the -current block and the mode that minimizes the differences between the original and predicted blocks is selected. Figure 11.2 shows the nine prediction modes used in the AIC codec. To save bits on coding the prediction mode, the blocks in the Cb and Cr channels use the same prediction modes as their corresponding blocks in the Y channel. The predicted block is subtracted from the original block to form a residual block. When a good prediction mode is chosen, the values in this residual block are smaller than the original pixel values and are, thus, better compressible.

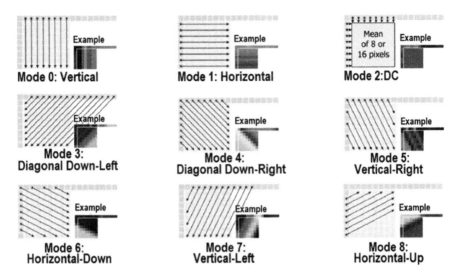

Figure 11.2   Prediction modes used by the AIC codec.

To further reduce the entropy of the residual block, this block is transformed using the DCT. The resulting transform coefficients are usually smaller than the original values. By quantizing these coefficients with a certain value based on a chosen quality level, the coefficients are reduced even further. As a result, many of the coefficients in the block will be zero, and most others will be close to zero. The prediction mode and quantized DCT coefficients are encoded to the stream using a binary arithmetic coding scheme that adapts to the context in which these values are coded.

This activity is a work in progress and it is continuously updated to accommodate new coding architectures and evaluation methodologies.

12

JPEG LS

JPEG LS (JPEG, Workplan & Specs of JPEG-LS, 2018) is the new lossless/ near-lossless compression standard for continuous-tone images, ISO-14495-1/ITU-T.87. The standard is based on the LOCO-I algorithm (LOw COmplexity LOssless COmpression for Images) developed at Hewlett-Packard Laboratories (Weinberger & Seroussi, 2000). LOCO-I attains compression ratios similar or superior to those obtained with schemes based on arithmetic coding.

JPEG LS was defined to address the needs for effective lossless and near-lossless compression of continuous-tone still images. This standard can be broken into two parts: ISO/IEC 14495-1:1999|ITU-T Rec. T.87 (1998), defining the core technology and ISO/IEC 14495-2:2003|ITU-T Rec. T.870 (03/2002), containing the extensions. JPEG-LS is especially suited for low-complexity hardware implementations of very moderate complexity, while, at the same time, providing state-of-the-art lossless compression (JPEG, Overview of JPEG-LS, 2018), (JPEG, Workplan & Specs of JPEG-LS, 2018), (JPEG, JPEG-LS Software, 2018).

12.1 JPEG LS Parts

JPEG LS currently includes the following parts:

Part 1. Baseline: Specifies the core coding system of JPEG LS.
Part 2. Extensions: Specifies extensions to the JPEP LS core coding system. It includes additional lossless multi-component transformations, an arithmetic coding option, and other improvements of the entropy coding mechanism. It also provides options for tuning near-lossless coding for optimal visual performance.

The JPEG LS syntax supports interleaved and non-interleaved (i.e., component by component) modes. In interleaved modes, possible correlation between color planes is used in a limited way. For some color spaces (e.g., an RGB

235

representation), good decorrelation can be obtained through simple lossless color transforms as a pre-processing step to JPEG LS. Similar performance is attained by more elaborate schemes that do not assume prior knowledge of the color space (Wu, Choi, & Memon, 1998).

JPEG LS offers a "near-lossless" mode which is a lossy mode of operation in which every sample value, in a reconstructed image component, differs from the corresponding original value in the original image by up to a preset (small) amount, δ.

JPEG LS is a "low complexity projection" based on a simple fixed context model, matching its modeling unit to a simple coding unit. Lossless image coding schemes often consist of two distinct and independent parts: modeling and coding. In the first part, one observes an image pixel by pixel (e.g., in raster scan). After observing i samples, we want to make inferences on the $i + 1$ pixel, $x^{i+1} = x^1...x^i$ by assigning a conditional probability distribution $P(\bullet \,|\, x^i)$ to the $i+1$ pixel (this distribution, learned from the past in a sequential formulation, is available at the decoder side as it decodes the sample pixels; in a two-pass scheme, the conditional distribution can be learned from the entire image in a first pass and sent to the decoder as side information), with an ideal contributed code by x^{i+1} of $-\log_2 p(x^{i+1}|x^i)$ bits which averages to the entropy of the probabilistic model (Weinberger & Seroussi, 2000).

Prior knowledge leads to a more accurate estimation. Hence, prior knowledge of the structure of the input image can be useful by adapting parametric distributions with few parameters per context to the data. Also, it is widely accepted that prediction residuals in continuous-tone images follow a two-sided geometric distribution (TSGD) centered at zero (Netravali & Limb, 1980).

Lossless image compression assigns probability according to the following components (Weinberger, Seroussi, & Saprio, 2000):

1. The predicted sample $\hat{x}_{i+1}$ is guessed from the next sample x_{i+1} based on the past data available.
2. Determination of the context in which x_{i+1} occurs.
3. A probabilistic model for the prediction residual $\varepsilon_{i+1} \triangleq x_{i+1} - \hat{x}_{i+1}$ given the context in which x_{i+1} occurs.

Note that the probability of an integer is expressed as $P_\theta(\varepsilon) \propto \theta^{|\varepsilon|}, \ \theta \in (0,1)$.

12.2 JPEG-LS Encoder Architecture

Figure 12.1 shows the block diagram of the JPEG-LS encoder (Rane & Sapiro, 2001). The diagram shows two main parts, the modeler and the coder. The

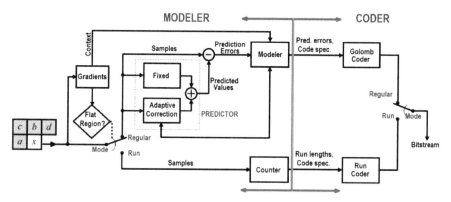

Figure 12.1 JPEG-LS block diagram.

modeler consists of the following modules: A decision based on the gradients of the samples, a predictor, a context modeler, and the run counter.

12.2.1 JPEG-LS Modeler

Context: The context modeling uses the causal template a, b, c, and d depicted in Figure 12.1. When encoding the first line of a source image component, the samples at positions b, c, and d are not present, and their reconstructed values are defined to be zero. If the sample at position x is at the beginning or at the end of a line, so that either a and c, or d is not present, the reconstructed value for a sample in position a or d is defined to be equal to Rb, the reconstructed value of the sample at position b, or zero for the first line in the component. The reconstructed value at a sample in position c, in turn, is copied (for lines other than the first line) from the value that was assigned to Ra when encoding the first sample in the previous line.

After a number of samples have been coded scanning from left to right and from top to bottom, the sample x depicted in Figure 12.1 shall be encoded. The context at this sample shall be determined by the previously reconstructed values Ra, Rb, Rc, and Rd corresponding to the samples a, b, c, and d, respectively.

The context at the encoder side is built out of the local gradients $g_1 = d - b$, $g_2 = b - c$ and $g_3 = c - a$; these differences capture the activity of the surrounding sample (smooth or edge) and affect the model in the same way. The differences are mapped by a quantizer $\kappa(\bullet)$ to a small number, assuming connected regions, and the same uniform probability in order to maximize the mutual information between the current sample value and its context.

To keep a low complexity, a fixed number of equiprobable regions are assumed. To preserve symmetry, the regions are labeled as $-T,\ldots,-1,0,1,\ldots,T$ and $\kappa(g) = -\kappa(-g)$ computing a total of $(2T + 1)^3$ different contexts. A further reduction in the number of contexts is obtained after observing that, by symmetry. JPEG-LS uses a total of $\left((2T+1)^3 +1\right)/2$ contexts and $T = 4$, resulting in 365 contexts. This number balances storage requirements (which are roughly proportional to the number of contexts) with high-order conditioning. A larger number of contexts result in a marginal compression gain and an excessive number of resources.

For an 8-bit/sample alphabet, the default quantization regions are $\{0\}$, $\pm\{1,2\}, \pm\{3,4,5,6\}, \pm\{7,8,\ldots,20\}, \pm\{c|c\geq 21\}$. The boundaries are adjustable parameters, except that the central region must be $\{0\}$. In particular, a suitable choice collapses quantization regions, resulting in a smaller effective number of contexts, with applications to the compression of small images. For example, a model with 63 contexts ($T = 2$) was found to work best for the 64×64-tile size (Digital-Imaging-Group, 1997).

When the differences g_1, g_2, g_3 are equal or less than a predetermined amount P_d the region is considered as flat region. In other words, the run mode is selected when it is estimated from the context that successive samples are very likely to be nearly identical within the tolerances required for near-lossless coding (identical, for lossless coding). The regular mode is selected when it is estimated from the context that samples are not very likely to be nearly identical within the tolerances required for near-lossless coding (identical, for lossless coding).

Predictor: Predictor is context dependent, follows an autoregressive (AR) model, and is used to reduce the number of coding parameters needed to model high-order dependencies because of the multiple conditional distributions with similar shape and different mean. Predictor includes a fixed predictor to detect vertical or horizontal edges – to incorporate prior knowledge by selecting among three simple predictors – and an adaptive correction module, also known as "bias cancellation," that is limited to an integer additive term, which is context-dependent (Weinberger, Rissanen, & Arps, 1996). Ideally, the value guessed for the current sample should depend on an adaptive model. The three fixed predictors are

$$\hat{x}_{\mathrm{MED}} = \begin{cases} \min(a,b) & \text{if } c \geq \max(a,b) \\ \max(a,b) & \text{if } c \leq \min(a,b) . \\ a+b-c & \text{otherwise} \end{cases} \qquad (12.1)$$

Predictor of Equation (12.1) picks b in case of a vertical edge is at left of the current location, a in case of a horizontal edge is above the current location, or $a + b - c$ if no edge is detected (Martucci, 1990). The guessed value can be seen as the median of the fixed values a, b, and $a + b - c$. This predictor was renamed during the standardization process "median edge detector" (MED).

Example 12.1:

60	105	100	105
50	**100**	**102**	60

Original image
(bold values are under prediction)

60	105	100	105
50	**95**	**100**	**105**

Predictive values
(blue values are the predicted values)

Differences e = { 5, 2, –45}.

The input pixel d is used in the adaptive correction block as part of the estimation procedure for the TSGD. However, it was observed that a DC offset is present in context conditioned prediction error signals due to integer-value constraints and possible bias in the prediction step. Therefore, a model for a TSGD should include a non-integer offset (μ). The fixed prediction offset is broken into two parts: an integer part R (or "bias") and a fractional part $s \in [0,1)$ (or "shift"); therefore, $\mu = R - s$. Hence, the probability distribution model assumed by the LOCO-I/JPEG-LS for the residuals of fixed predictors at each context is

$$P_{(\theta,\mu)}(\varepsilon) = \frac{1-\theta}{\theta^{1-s}+\theta^s}\theta^{|\varepsilon-R+s|}, \quad \theta \in (0,1), \quad \varepsilon = 0,\pm1,\pm2,.... \qquad (12.2)$$

The bias calls for an integer adaptive term in the predictor. Also, it is assumed that this term is tuned to cancel R, producing average residuals between distribution modes located at 0 and –1. Therefore, Equation (12.2) can be reduced to

$$P_{(\theta,\mu)}(\varepsilon) = \frac{1-\theta}{\theta^{1-s}+\theta^s}\theta^{|\varepsilon+s|}, \quad \theta \in (0,1), \quad s \in [0,1), \varepsilon = 0,\pm1,\pm2,.... \qquad (12.3)$$

When $s = 1/2$, $P_{(\theta,\mu)}$ is a bi-modal distribution with equal peaks at and 0. Note that the parameters of the distribution are context-dependent (Merhav, Seroussi, & Weinberger, 2000). The image alphabet has a finite size α. However, in practice, for a given prediction $\hat{x}$, ε is in the range $-\hat{x} \le \varepsilon \le \alpha - \hat{x}$. Since $\hat{x}$ is also known at the decoder side, prediction residual can be reduced modulo α, to a value in the range $-\lfloor \alpha/2 \rfloor$ and $\lfloor \alpha/2 \rfloor - 1$. This remaps large prediction residuals to small ones.

The adaptive part of the predictor is context-based and it cancels R of the offset due to the fixed predictor. An estimate of R based on the average can be obtained by just keeping a count N of context occurrences, and a cumulative sum D of fixed prediction errors incurred so far in the context. Then, a correction value C' can be computed as the rounded average $C' = \lceil D/N \rceil$ and added to the prediction $\hat{x}_{MED}$. However, the average requires a division and it is very sensitive to outliers. Hence, the following equivalent is used:

$$D = NC' + B', \tag{12.4}$$

where B' is an integer that satisfies $-N < B' \leq 0$. It can be implemented by storing B' and C' and adjusting both for each occurrence of the context. The corrected prediction error ε is first added to B' and then N is subtracted or added until the result is in the desired range $(-N,0]$. The number of subtractions/additions prescribes the adjustment for C'.

The above procedure is further modified in LOCO-I/JPEG-LS by limiting the number of subtractions/additions to one per update, and then "forcing" the value of B' into the desired range, if needed. Specifically, B' and C' are replaced by approximate values B and C, which are initialized to zero and updated according to the following division-free procedure (Weinberger, Seroussi, & Saprio, 2000):

$$B = B + \varepsilon;$$
$$N = N + 1;$$
$$\text{if } (B \leq -N)$$
$$\{ \quad C = C - 1;$$
$$B = B + N;$$
$$\text{if } (B \leq -N) \; B = -N + 1;$$
$$\}$$
$$\text{else if } (B > 0)$$
$$\{ \quad C = C + 1;$$
$$B = B - N;$$
$$\text{if } (B > 0) \; B = 0;$$
$$\}$$

Figure 12.2   Division-free for bias computation procedure.

This procedure will tend to produce average prediction residuals in the interval $(-1,0]$, with C being an estimate of R.

12.2.2 JPEG-LS Coder

Run mode: If the reconstructed values of the samples at a, b, c, and d are identical for lossless coding, or the differences between them (the local gradients – see Example 12.1) are within the bounds set for near-lossless coding, the context modeling procedure selects the run mode, and the encoding process skips the prediction and error encoding procedures.

In run mode, the encoder looks, starting at x, for a sequence of consecutive samples with values identical (or within the bound specified for near-lossless coding) to the reconstructed value of the sample at a. A run is ended by a sample of a different value (or one which exceeds the bound specified for near-lossless coding), or by the end of the current line, whichever comes first. The length information, which also specifies one of the above two run-ending alternatives, is encoded using a procedure which is extended from Golomb coding but has improved performance and adaptability (ISO/IEC-14495, 1999).

Regular mode (prediction and error encoding): In the regular mode, the context determination procedure is followed by a prediction procedure. The predictor combines the reconstructed values of the three neighborhood samples at positions a, b, and c to form a predicted sample value at position x as shown in Figure 12.1. The prediction error is computed as the difference between the actual sample value at position x and its predicted value. This prediction error is then corrected by a context-dependent term to compensate for systematic biases in prediction. In the case of near-lossless coding, the prediction error is quantized.

The corrected prediction error (further quantized for near-lossless coding) is then encoded using a procedure derived from Golomb coding. Note that Golomb coding corresponds to Huffman coding for a geometric distribution.

The compressed image data output by the encoding process consists of marker segments and coded image data segments. The marker segments contain information required by the decoding process, including the image dimensions. The interchange format is the coded representation of compressed image data for exchange between application environments. Figure 12.3 shows the (a) preset parameters marker segment syntax and (b) preset coding parameters.

LSE (X'FFF8'): Preset parameters marker; marks the beginning of the JPEG-LS preset parameters marker segment.

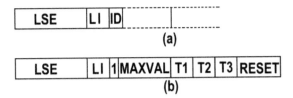

Figure 12.3   LSE marker (a) preset parameters marker segment syntax and (b) preset coding parameters.

LI: Preset parameters length; specifies the length of the JPEG-LS preset parameters marker segment.

ID: Specifies which JPEG-LS preset parameters follow. If ID = X'01', the JPEG-LS preset coding parameters follow. If ID = X'02', a mapping table specification follows. If ID = X'03', a mapping table continuation follows. If ID = X'04', X and Y parameters greater than 16 bits are defined.

MAXVAL: Maximum possible value for any image sample in the scan. This must be greater than or equal to the actual maximum value for the components in the scan.

T1: First quantization threshold value for the local gradients.

T2: Second quantization threshold value for the local gradients.

T3: Third quantization threshold value for the local gradients.

RESET: Value at which the counters A, B, and N are halved.

Note: If the local gradients are all zero (for lossless coding), or their absolute values are less than or equal to the parameter NEAR, the allowed error for near-lossless coding, the encoder shall enter the run mode; otherwise, the encoder shall enter the regular mode.

Coding parameters: The bits generated by the encoding process forming the compressed image data shall be packed into 8-bit bytes. These bits shall fill bytes in decreasing order of significance. As an example, when outputting a binary code a_n, a_{n-1}, a_{n-2}, ..., a_0, where a_n is the first output bit, and a_0 is the last output bit, a_n will fill the most significant available bit position in the currently incomplete output byte, followed by a_{n-1}, a_{n-2}, and so on. Then, when an output byte is completed, it is placed as the next byte of the encoded bit stream, and a new byte is started. An incomplete byte, just before a marker, is padded with zero-valued bits before the insertion of any marker.

Marker segments are inserted in the data stream. In order to provide for easy detection of marker segments, a single byte with the value X'FF' in a coded image data segment shall be followed with the insertion of a single bit "0." This inserted bit shall occupy the most significant bit of the next byte.

If the X'FF' byte is followed by a single bit "1," then the decoder processes the byte which follows as the second byte of a marker and processes it. If a "0" bit was inserted by the encoder, the decoder shall discard the inserted bit, which does not form part of the data stream to be decoded.

Codec initialization: The following initializations shall be performed at the start of the encoding process of a scan:

a. Compute the parameter RANGE: For lossless coding (NEAR = 0), RANGE = MAXVAL + 1. For near-lossless coding (NEAR > 0),

$$RANGE = \left| \frac{MAXVAL + 2 \times NEAR}{2 \times NEAR + 1} \right| + 1.$$

Compute $qbpp = \log(RANGE)$, $bpp = \max\left(2, \lceil \log(\max val + 1) \rceil\right)$, $LIMIT = 2 \times (bpp + \max(8, bpp))$

b. Initialize the variables $N[0..366]$, $A[0..366]$, $B[0..364]$, and $C[0..364]$, where the nomenclature $[0...i]$ indicates that there are $i + 1$ instances of the variable. The instances are indexed by $[Q]$, where Q is an integer between 0 and i. The indexes $[0...364]$ correspond to the regular mode contexts (total of 365), while the indexes $[365]$ and $[366]$ correspond to run mode interruption contexts; i.e., $C[5]$ corresponds to the variable C in the regular mode context indexed by 5. Each one of the entries of A is initialized with the value

$$\max\left(2, \left| \frac{RANGE + 2^5}{2^6} \right| \right).$$

Those of N are initialized with the value 1, and those of B and C with the value 0.
c. Initialize the variables for the run mode: RUNindex = 0 and J[0...31] = {0, 0, 0, 0, 1, 1, 1, 1, 2, 2, 2, 2, 3, 3, 3, 3, 4, 4, 5, 5, 6, 6, 7, 7, 8, 9, 10, 11, 12, 13, 14, 15}.
d. Initialize the two run interruption variables $Nn[365]$ and $Nn[366]$ to 0.
e. Set current sample to the first sample in the source image.

Encoding procedure for single component: In order to provide for easy detection of marker segments, a single byte with the value X'FF' in a coded image data segment shall be followed with the insertion of a single bit "0."

This inserted bit shall occupy the most significant bit of the next byte. If the X'FF' byte is followed by a single bit "1," then the decoder shall treat the byte which follows as the second byte of a marker and process it. If a "0" bit was inserted by the encoder, the decoder discards the inserted bit, which does not form part of the data stream to be decoded. The encoding procedure is summarized as follows.

12.2.3 Encoding Procedure for Single Component

1. Initialization:
    a. Assign default parameter values to JPEG-LS preset coding parameters not specified by the application (see the "Coding parameters" section).
    b. Initialize the non-defined samples of the causal template (see the "Codec initialization" section).
    c. Compute the parameter RANGE and the parameters qbpp, bpp, and LIMIT (see the "Codec initialization" section).
    a. For each context Q, initialize four variables:

    $$A[Q] = \max\left(2, \left\lfloor \frac{RANGE + 2^5}{2^6} \right\rfloor\right), \quad B[Q] = C[Q] = 0, \quad N[Q] = 1. \text{ For}$$

    $A[Q]$ and $N[Q]$, Q is an integer between 0 and 366; for $B[Q]$ and $C[Q]$, Q is an integer between 0 and 364 (regular mode contexts only).
    d. Initialize RUNindex $= 0$ and $J[0...31]$ (see the "Codec initialization" section).
    e. Initialize the two run interruption variables $Nn[365]$ and $Nn[366]$ to 0.
    f. Set current sample to the first sample in the source image.
2. For the current sample, compute the local gradients.
3. Select the coding mode. If run mode is selected, go to Step 17; otherwise, continue with the regular mode.
4. Quantize the local gradients according to the following portion of C code:

```
if (g_i <= -T3)          Qi = -4;
else if (g_i <= -T2)     Qi = -3;
else if (g_i <= -T1)     Qi = -2;
else if (g_i < - NEAR)   Qi = -1;
else if (g_i <= NEAR)    Qi = 0;
else if (g_i < T1)       Qi = 1;
```

Table 12.1 LSE marker segment parameter sizes and values for JPEG-LS preset coding parameters.

Parameter	Size (bits)	Values
Ll	16	13
ID	8	1
MAXVAL	16	0, or $1 \leq MAXVAL < 2^P$
T1	16	0, or $NEAR + 1 \leq T1 \leq MAXVAL$
T2	16	0, or $T1 \leq T2 \leq MAXVAL$
T3	16	0, or $T2 \leq T3 \leq MAXVAL$
RESET	16	0, or $3 \leq RESET \leq max(255, MAXVAL)$

$$\text{else if } (g_i < T2) \quad Qi = 2;$$
$$\text{else if } (g_i < T3) \quad Qi = 3;$$
$$\text{else } Qi = 4;$$

NEAR is the difference bound for near-lossless coding (see Table 12.1).

P is the number of bits per image sample, contained in the start of frame marker segment.

For MAXVAL, T1, T2, T3, and RESET, a value of 0 indicates reverting to default values as $MAXVAL = 2^P - 1$, RESET = 64. The default threshold values T1, T2, and T3, for gradient quantization, are given in terms of MAXVAL, NEAR, and the "basic" default threshold values for the case MAXVAL = 255, lossless coding (NEAR = 0), denoted BASIC_T1 = 3, BASIC_T2 = 7, and BASIC_T3 = 21. Using a clamping function for integers i and j as

CLAMP(i,j)
   if ($i >$ MAXVAL or $i < j$) return j;
   else return I;

In the case where MAXVAL $\geq$ 128, the dependence of the default values on MAXVAL is specified by FACTOR = floor ((min (MAXVAL, 4095) + 128)/256). The default values in this case are given by

T1 = CLAMP(FACTOR * (BASIC_T1 − 2) + 2 + 3*NEAR, NEAR + 1)
T2 = CLAMP(FACTOR * (BASIC_T2 − 3) + 3 + 5*NEAR, T1)
T3 = CLAMP(FACTOR * (BASIC_T3 − 4) + 4 + 7*NEAR, T2).

Otherwise, if MAXVAL < 128, the dependence of the default values on MAXVAL is specified by FACTOR = floor (256/(MAXVAL + 1)). The default values in this case are given by
T1 = CLAMP(max(2, BASIC_T1/FACTOR + 3*NEAR), NEAR + 1)

T2 = CLAMP(max(3, BASIC_T2/FACTOR + 5*NEAR), T1)
T3 = CLAMP(max(4, BASIC_T3/FACTOR + 7*NEAR), T2).

5. Check and change, if necessary, the signs of the components of the vector representing the context, modifying accordingly the variable SIGN.

   If the first non-zero element of the vector $(Q1, Q2, Q3)$ is negative, then all the signs of the vector $(Q1, Q2, Q3)$ shall be reversed to obtain $(-Q1, -Q2, -Q3)$. In this case, the variable SIGN shall be set to -1; otherwise, it shall be set to $+1$. After this possible "merging," the vector $(Q1, Q2, Q3)$ is mapped, on a one-to-one basis, into an integer Q representing the context for the sample x.

   The function mapping the vector $(Q1, Q2, Q3)$ to the integer Q is not specified in this Recommendation|International Standard. This Recommendation requires that the mapping shall be one-to-one, that it shall produce an integer in the range $[0...364]$, and that it be defined for all possible values of the vector $(Q1, Q2, Q3)$, including the vector $(0, 0, 0)$.

6. Compute the predicted value of sample x according to Equation (12.1).
7. Correct the predicted values of x using $C[Q]$ and the variable SIGN, and clamp the corrected value to the interval $[0 ... MAXVAL]$ according to the following C code:

   if (SIGN == 1) Px = Px + C[Q];
   else Px = Px − C[Q];

   if (Px > MAXVAL) Px = MAXVAL;
   else if (Px < 0) Px = 0;

8. Compute the prediction error and, if necessary, invert its sign as

   Errval = Ix − Px;
   if (SIGN == −1) Errval = − Errval

9. For near-lossless coding, quantize the error and compute the reconstructed value of the current sample as follows:

   if (Errval > 0) Errval = (Errval + NEAR) / (2 * NEAR + 1);
   else Errval = − (NEAR − Errval)/(2 * NEAR + 1);

   Rx = Px + SIGN * Errval * (2 * NEAR + 1);
   if (Rx < 0)
       Rx = 0;
   else if (Rx > MAXVAL)
       Rx = MAXVAL;

For lossless coding, update the reconstructed value by setting Rx equal to Ix.

10. Reduce the error to the relevant range as

> if (Errval < 0) Errval = Errval + RANGE;

> if (Errval >= ((RANGE + 1)/2)) Errval = Errval – RANGE;

11. Compute the context-dependent Golomb variable *k* as

> for (*k* = 0; (*N*[*Q*]<<k)<*A*[*Q*]; *k*++);

12. Perform the error mapping to non-negative values

> if ((NEAR == 0) && (*k* == 0) && (2 * *B*[*Q*] <= – *N*[*Q*]))
> {
>   if (Errval >= 0) MErrval = 2 * Errval + 1
>   else MErrval = –2 * (Errval + 1);
> }
> else
> {
>   if (Errval >= 0) MErrval = 2 * Errval;
>   else MErrval = –2 * Errval – 1;
> }

13. Encode the mapped error value MErrval using the limited length Golomb code function LG (*k*, LIMIT) defined by the following procedure:

   1. If the number formed by the high-order bits of MErrval (all but the *k* least significant bits) is less than LIMIT – qbpp – 1, this number shall be appended to the encoded bit stream in unary representation, i.e., by as many zeros as the value of this number, followed by a binary one. The *k* least significant bits of MErrval shall then be appended to the encoded bit stream without change, with the most significant bit first, followed by the remaining bits in decreasing order of significance.

   2. Otherwise, LIMIT – qbpp – 1 zero shall be appended to the encoded bit stream, followed by a binary one. The binary representation of MErrval – 1 shall then be appended to the encoded bit stream using qbpp bits, with the most significant bit first, followed by the remaining bits in the decreasing order of significance.

14. Update the variables according to the following procedure:
    B[Q] = B[Q] + Errval *(2 *NEAR + 1);
    A[Q] = A[Q] + abs(Errval);

    if (N[Q] == RESET)
    {
            A[Q] == A[Q] >> 1;
            if (B[Q] >= 0) B[Q] = B[Q] >> 1;
            else B[Q] = –((1 – B[Q]) >> 1);
            N[Q] = N[Q] >> 1;
    }
    N[Q] = N[Q] + 1;

15. Update the prediction correction value C[Q] according to
    if (B[Q] <= –N[Q])
    {
            B[Q] = B[Q] + N[Q];
            if (C[Q] > MIN_C) C[Q] = C[Q – 1];
            if (B[Q] <= –N[Q]) B[Q] = –N[Q] + 1;
    }
    else if (B[Q] > 0)
    {
            B[Q] = B[Q] – N[Q];
            if C[Q] < MAX_C) C[Q] = C[Q] + 1;
            if (B[Q] > 0) B[Q] = 0;
    }

    Note: MAX_C is the maximum allowed value of C[0...364], equal to
    127. MIN_C is the minimum allowed value of C[0...364], equal to –128.

16. Go to Step 2 to process the next sample.
17. Run mode coding:
    a. Set RUNval = Ra. While (abs (Ix – RUNval) <= NEAR), increment
       RUNcnt, and if not at the end of a line, read a new sample. Set Rx =
       RUNval each time the sample *x* is added to the run according to
            RUNval = Ra;
            RUNcnt = 0;
            while (abs (Ix – RUNval) <= NEAR)
            {
                    RUNcnt = RUNcnt + 1;

```
            Rx = RUNval;
            if (EOLine == 1) break;
            else GetNextSample();
    }
```

Note: The test abs(Ix − RUNval) <= NEAR reduces, in the lossless case, to Ix == RUNval.

b. While RUNcnt ≥ $2^{J[RUNIndex]}$, do

```
        while (RUNcnt >= (1 << J[RUNindex]))
        {
            AppendToBitStream(1,1);
            RUNcnt = RUNcnt – (1 << J[RUNindex]);
            if (RUNindex < 31) RUNindex = RUNindex +1;
        }
```
Append "1" to the bit stream. RUNcnt = RUNcnt − $2^{J[RUNIndex]}$. If RUNindex < 31, then increment RUNindex by 1.

c. If the run was interrupted by the end of a line
```
        if (abs(Ix – RUNval) > NEAR)
        {
            AppendToBitStream(0,1);
            AppendToBitStream(RUNcnt, J[RUNindex]);
            if (RUNindex > 0) RUNindex = RUNindex –1;
        }
        else if (RUNcnt > 0) AppendToBitStream(1,1);
```

Note:
     i. If RUNcnt > 0, append "1" to the bit stream.
     ii. Go to Step 16.

d. Append "0" to the bit stream according to the *C* code segment in 17c.

e. Append RUNcnt in binary representation (using *J*[RUNindex] bits) to the bit stream (MSB first as in the *C* code segment 17c).

f. If RUNindex > 0, then decrement RUNindex by 1 (see *C* code segment 17c).

18. Run interruption sample encoding:

a. Compute the index RItype
```
        if (abs(Ra – Rb) <= NEAR ) RItype = 1;
```

else RItype = 0;

b. Compute the prediction error

    if (RItype ==1) Px = Ra;
    else Px = Rb;
    Errval = Ix – Px;

c. Correct, if necessary, the sign of Errval

    if ((RItype == 0) && (Ra > Rb))
    {
        Errval = –Errval;
        SIGN = –1;
    }
    else SIGN = 1;

    if (NEAR > 0)
    {
        Errval = Quantize(Errval);
        Rx = ComputeRx ();
    }
    else Rx = Ix;
    Errval = ModRange (Errval,RANGE);

This step is analogous to the context-merging procedure in the regular coding mode. For near-lossless coding, Errval shall be quantized and Rx computed as in the code of Step 9. The error shall then be reduced using the variable RANGE. This reduction is performed by the function ModRange shown in 18c.

d. Compute the auxiliary variable TEMP. This variable is used for the computation of the Golomb variable k.

    if (RItype == 0) TEMP = A[365];
    else TEMP = A[366] + (N[366] >> 1);

e. Set Q = RItype + 365. The Golomb variable k shall be computed, following the same procedure as in Step 11, as in the regular mode but using TEMP instead of $A[Q]$.

f. Compute the flag map as

    if ((k == 0) && (Errval > 0) && (2 * Nn[Q] < N[Q])) map = 1;
    else if ((Errval < 0) && (2 * Nn[Q] >= N[Q])) map = 1;
    else if ((Errval < 0) && (k ! = 0)) map = 1;
    else map = 0;

This variable influences the mapping of Errval to non-negative values, as

EMErrval = 2 * abs(Errval) – RItype – map;

g. Encode EMErrval following the same procedures as in the regular mode but using the limited length Golomb code function LG(k, glimit), where glimit = LIMIT – J[RUNindex] – 1 and RUNindex corresponds to the value of the variable before the decrement specified in the C function of Step 17c.

h. Update the variables for run interruption sample encoding as

if Errval < 0) Nn[Q] = Nn[Q] + 1;

A[Q] = A[Q] + ((EMErrval + 1 RItype) >> 1);
if (N[Q] == RESET)
    {
        A[Q] = A[Q] >> 1;
        N[Q] = N[Q] >> 1;
        Nn[Q] = Nn[Q] >> 1;
    }
    N[Q] = N[Q] + 1;
19. Go to Step 16.

For non-interleaved mode (Ns = 1, ILV = 0), the minimum coded unit is one line. For sample interleaved mode (Ns > 1, ILV = 2), the MCU is a set of Ns lines, one line per component, in the order specified in the scan header. The order in the scan header is determined by the order in the frame header.

For line interleaved mode (Ns > 1, ILV = 1), the MCU is V1 lines of component C1 followed by V2 lines of component C2 followed by VNs lines of component CNs. In addition, the encoding process shall extend the number of lines if necessary so that the last MCU is completed. Any line added by an encoding process to complete a last partial MCU shall be removed by the decoding process.

For encoding images with more than one component (e.g., color images), standard supports combinations of single-component scans (each component can be encoded completely and separately) and multi-component scans (JPEG, Workplan & Specs of JPEG-LS, 2018). Figure 12.4 shows an example of a multi-component image that contains three components. The larger the

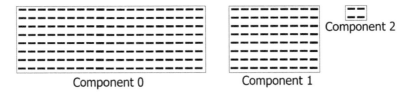

Figure 12.4   Example of sampling for a three-component image.

sampling factor, the more lines in that dimension for a particular component. For a red–green–blue color image, these sampling factors may be the same for each of the three components. In a Y-U-V color image, the luminance component, Y, may have higher sampling factors than the two chrominance components, U and V. (This representation exploits the relative spatial sensitivity to chrominance changes with respect to the luminance spatial sensitivity.) There are two ways a multi-component image can be interleaved: line interleaving and sample interleaving.

For multi-component scans, two modes (described below) are supported: line interleaved and sample interleaved. In the line interleaved mode, a predetermined number of lines from each component are encoded in a fashion very similar to the single-component mode. The data for the given number of lines in each component is encoded with the data for component 2 following the data for component 1 and the data for component 3 following the data for component 2, etc. The sample-interleaved mode, pixels from each component are encoded sequentially from one component to the next before the next pixel in the first component is encoded. In this case, all components in a scan must have the same dimension.

Figure 12.5 shows an example of the line interleaving for a given multi-component image example. In the example $V_0 = 4$, four lines are coded from component 0. $V_1 = 2$; so two lines of component 1 are coded (note that these are shorter lines). Finally, $V_2 = 1$ and a single line is coded from component 2. In line interleaving, the determination of run versus regular mode processing is identical to the process used for single-component images.

The specific components per scan are specified in the scan header, as well as the interleave mode (as specified by parameter ILV), which describes the structure within a single scan. The parameter ILV admits the values 0 (non-interleaved), 1 (line interleaved), and 2 (sample interleaved).

For multi-component scans, a single set of context counters (A, B, C, N, and Nn) is used across all the components in the scan. The prediction and context modeling procedures shall be performed as in the single-component case and are component independent, meaning that samples

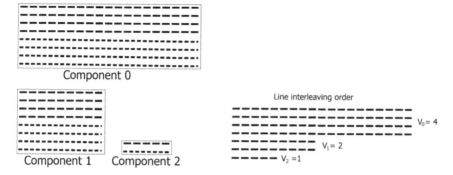

Figure 12.5 Example line interleaving for a three-component image.

from one component are not used to predict or compute the context of samples from another component. All the encoding and decoding variables (e.g., $A[0...366]$) shall be set to their initial values, when a new scan is to be encoded. The dimensions of each component are given by the information in the frame header. The byte completion padding applies also to multi-component scans.

12.3 Marker Assignment

Two markers from the JPGn set (which were reserved for JPEG extensions) are assigned.

SOF55 (X'FFF7') identifies a new Start of Frame marker for JPEG-LS processes.

LSE (X'FFF8') identifies marker segments used for JPEG-LS preset parameters.

Each of these markers starts a marker segment. These marker segments begin with a 2-byte segment length parameter. In addition, the SOI, EOI, SOS, DNL, DRI, RSTm, APPn, and COM are valid markers in JPEG-LS.

12.3.1 Decoding Procedure

1. Initialization:
   a. Assign default parameter values to JPEG-LS preset coding parameters not specified by the application (see the "Coding Parameters" section).
   b. Initialize the non-defined samples of the causal template (see the "Initialization" section).

c. Compute the parameter RANGE and the parameters qbpp, bpp, and LIMIT (see the "Initialization" section).

d. For   each   context Q,   initialize   four   variables:

$$A[Q] = \max\left(2, \left|\frac{\text{RANGE} + 2^5}{2^6}\right|\right), \quad B[Q] = C[Q] = 0, \quad N[Q] = 1.$$

For $A[Q]$ and $N[Q]$, Q is an integer between 0 and 366; for $B[Q]$ and $C[Q]$, Q is an integer between 0 and 364 (regular mode contexts only).

e. Initialize $RUNindex = 0$ and $\mathbf{J}[0\ldots31]$.

f. Initialize the two run interruption variables $Nn[365]$ and $Nn[366]$ to 0.

2. Compute the local gradients.

3. Select the coding mode. If run mode is selected, go to Step 18; otherwise, continue with the regular mode.

4. Quantize the local gradients according to the C code of Step 4 in the "Encoding procedure" section.

5. Check and change if necessary the sign of the context, modifying accordingly the variable SIGN by using to the C code of Step 5 in the "Encoding procedure" section.

6. Compute the predicted value of sample x according to Equation (12.1).

7. Correct Px using $C[Q]$ and the variable SIGN and clamp the corrected value to the interval $[0\ldots\text{MAXVAL}]$ according to the procedure of Step 7 in the "Encoding procedure" section.

8. Compute the context-dependent Golomb variable k according to the procedure: for $(k = 0; (N[Q]<<k)<A[Q]; k++)$;

9. Decode the mapped error value MErrval:

   a. Read the unary code. If it contains less than LIMIT – qbpp – 1 zeros, use it to form the most significant bits of MErrval and read k additional bits, to compose the k least significant bits of MErrval.

   b. If the unary code contains LIMIT – qbpp – 1 zeros, read qbpp additional bits to get a binary representation of MErrval –1.

10. Perform the inverse of the error mapping indicated in the C code of Step 12 in the "Encoding procedure" section, where now MErrval is given and Errval is computed.

11. Update the variables according to C code of Step 14 in the "Encoding procedure" section.

12. For near-lossless coding, multiply Errval by $(2*\text{NEAR} + 1)$.

13. Invert sign of Errval if the variable SIGN is negative.

14. Compute Rx = (Errval + Px) modulo [RANGE*(2*NEAR + 1)]. For near-lossless coding, map Rx to the interval [–NEAR...RANGE*(2*NEAR + 1) – 1 – NEAR]. Clamp Rx to [0...MAXVAL]. This is done using the following *C* code:

    if (Rx <– NEAR) Rx = Rx + RANGE*(2*near + 1);
    else if (Rx > MAXVAL + NEAR) Rx = Rx – RANGE*(2*near + 1);
    if (Rx < 0) Rx = 0;
    else if (Rx > MAXVAL) Rx = MAXVAL;

15. Map Rx using the inverse point transform Pt specified by the parameter Al.
16. Compute the prediction correction value $C[Q]$ according to the update of bias-related variables $B[Q]$ and $C[Q]$ *C* code of Step 14 in the "Encoding procedure" section.
17. Process next sample of the same component or of next component, starting from Step 2.
18. Run mode decoding:
    a. If R = "1," then
        i. Fill the image with $2^{J[RUNindex]}$ samples of value R_a or until a line is completed. If exactly $2^{J[RUNindex]}$ samples were filled in the previous step, and RUNindex < 31, then increase RUNindex by 1.
        ii. If the last sample in the line has not yet been decoded, return to Step 18a to read more bits from the bit stream. Otherwise, go to Step 17.
    b. If R = "0," then
        i. Read $J[RUNindex]$ bits from the bit stream and fill the image with the value R_a for as many samples as the number formed by these bits (MSB first).
        ii. If RUNindex > 0, decrement RUNindex by 1.
        iii. Decode the run interruption sample value.
        iv. Go to Step 17.

13

JPEG XL

The JPEG XL Image Coding System (ISO/IEC 18181), (JPEG, Overview of JPEG XL, 2019) has a richer feature set than existing codecs and can deliver images with similar quality at a third of the size of widely used alternatives. It is designed with responsive web design in mind so that content renders well on a wide range of devices. The JPEG XL coding tools include variable-size DCT, nonlinear Haar transforms, multiresolution encoding, adaptive quantization, adaptive loop filters, and context modeling (JPEG, Overview of JPEG XL, 2019).

JPEG XL includes several features that help transition from the legacy JPEG format. Existing JPEG files can be losslessly transcoded to JPEG XL, while significantly reducing their size. A lightweight lossless conversion process back to JPEG ensures compatibility with existing JPEG-only clients such as older generation phones and browsers. Thus, it is easy to migrate to JPEG XL because servers can store a single JPEG XL file to serve both JPEG and JPEG XL clients. JPEG XL decoders can perform enhancement that would improve image quality when dealing with the legacy JPEG format. JPEG XL encoders may also choose to add a small amount of additional information to further enhance the quality of decoded images, while remaining backward compatible with existing legacy JPEG decoders.

JPEG XL is also designed to meet the needs of high-quality imaging and professional photography. A color-managed processing pipeline with full 32 bits per channel precision enables support for wide-color-gamut/ high-dynamic-range images. JPEG XL reaches high compression efficiency at visually lossless quality (as defined in ISO/IEC 29170-2) using psycho-visual modeling plugins.

JPEG XL is designed for efficient decoding even in software, with parallel and SIMD-friendly coding tools. JPEG XL compares favorably with contemporary coding solutions in terms of complexity.

JPEG XL further includes features such as animations, alpha channels, lossless, and progressive coding to support a wide range of use cases including, but not limited to, photo galleries, e-commerce, social media, user interfaces, and cloud storage. To enable novel applications, it also adds support for 360° images, image bursts, large panoramas/mosaics, and printing (JPEG, Overview of JPEG XL, 2019).

The JPEG Committee launched the Next-Generation Image Compression activity also known as JPEG-XL. This activity aims to develop a standard for image compression that offers substantially better compression efficiency than existing image formats (e.g., >60% over JPEG-1), along with features desirable for web distribution and efficient compression of high-quality images.

The document is the final Call for Proposals (CfP) for a Next-Generation Image Coding Standard and was issued as outcome of the 79th JPEG meeting, La Jolla, USA, April 9–15, 2018. The deadline for expression of interest and registration is August 15, 2018. Submissions to the Call for Proposals are due on September 1, 2018 (ISO/IEC--JTC1/SC29/WG1, 2018).

13.1 Scope

The next-generation image coding activity aims to develop an image coding standard that offers:
1. Significant compression efficiency improvement over coding standards in common use at equivalent subjective quality, e.g., >60% over JPEG.
2. Features for web applications, such as support for alpha channel coding and animated image sequences.
3. Support of high-quality image compression, including higher resolution, higher bit depth, higher dynamic range, and wider color gamut coding.

13.2 Use Cases

1. *Image-rich UIs and web pages on bandwidth-constrained connections:* Web sites and user interfaces become more and more image-driven. Images play a major role in the interaction between users, the selection of topics, stories, movies, articles, and so on. In these UIs, formats are preferred that are widely supported in browsers and/or CE devices, such as JPEG, PNG, and WebP. The intended applications are:
   a. social media;
   b. media distribution;
   c. cloud storage;
   d. media web sites;
   e. animated image.
2. *High-quality imaging applications:* On the high end, UIs utilize images that have larger resolutions and higher bit depths, and the availability

of higher dynamic range and wider color gamut is a benefit for vivid color imagery. 4K TVs are becoming mainstream, and HDR/WCG technology is picking up, leading to a shift to high-quality UIs. A new standard should provide efficient compression and high visual quality for these applications. Images in these applications can contain a mixture of natural images and synthetic elements (overlays, multilingual text, gradients, etc.). Some applications in this case are:

  a. HDR/WCG user interfaces;

  b. augmented/virtual reality.

13.3 Requirements

This CfP targets image coding technology that can at least support images with the following attributes:

- Image resolution: from thumbnail-size images up to at least 40 MP images.
- Transfer functions including those listed in BT. 709 (ITU-R-BT.709-6, 2015) and BT. 2100 (ITU-T-Rec-BT.2100, 2017).
- Bit depth: 8-bit and 10-bit.
- Color space: at least RGB, YCbCr, ICtCp.
  - ○ Input type of the encoder shall match output type of the decoder.
  - ○ Internal color space conversion is permitted (as part of the proposal).
- Color primaries including BT. 709 and BT. 2100 (ITU-R-BT.709-6, 2015), (ITU-T-Rec-BT.2100, 2017).
- Chrominance subsampling (where applicable): 4:0:0, 4:2:0, 4:2:2, and 4:4:4.
- Different types of content, including natural, synthetic, and screen content.

A desirable attribute for submitted technology is the support of up to 12 bits for nonlinear images and up to 16 bits for linear images.

13.4 Compressed Bit-Stream Requirements

Submissions shall cover at least the core requirements, and are encouraged to cover desirable requirements as well.

Core requirements
Significant compression efficiency improvement over coding standards in common use at equivalent subjective quality.
Hardware/software implementation-friendly encoding and decoding (in terms of parallelization, memory, complexity, and power consumption).

Support for alpha channel/transparency coding.
Support for animation image sequences.
Support for 8-bit and 10-bit bit depth.
Support for high dynamic range coding.
Support for wide color gamut coding.
Support for efficient coding of images with text and graphics.
Desirable requirements
Support for higher bit depth (e.g., 12- to 16-bit integer or floating-point HDR) images.
Support for different color representations, including Rec. BT.709, Rec. BT.2020, Rec. BT.2100, LogC.
Support for embedded preview images.
Support for very low file size image coding (e.g., < 200 bytes for 64 × 64 pixel image) (Cabral & Kandrot, 2015).
Support for lossless alpha channel coding.
Support for a low-complexity profile.
Support for region-of-interest coding.

13.5 Test Material

Test images include natural (color and gray-scale), computer generated and screen captured content, and HDR/WCG images. The classification of material is as follows:

1. Class A: natural images (color): 8-bit 4:4:4 and 10-bit.
2. Class B: natural images (gray-scale): 8-bit and 12-bit.
3. Class C: computer-generated images: 8-bit, 10-bit and 12-bit.
4. Class D: screen content images: 8-bit.
5. Class E: HDR/WCG images: 16-bit floating point (IEEE 754), 10-bit 1080p, 10-bit 4K.
6. Class F: natural images with overlays (text, logos, etc.).

All test material is available to proponents on an FTP server as per email request for the purpose of the standardization project only.

13.6 Anchors

The proposed methods must be compared against the following anchors:
- JPEG (ISO/IEC-10918, 1994);
- JPEG 2000 (ISO/IEC-15444, 2004);

- HEVC (ISO-23008-2-ITU-T-Rec.H.265, 2018);
- WebP (Google, 2019).

13.7 Target Rates

Target bit rates for the objective evaluations include 0.06, 0.12, 0.25, 0.50, 0.75, 1.00, 1.50, and 2.00 bpp. Target bit rates for the subjective evaluations will be a subset of the target bit rates for the objective evaluations and will depend on the complexity of the test images.

13.8 Objective Quality Testing

Objective quality testing must be done by computing several quality metrics, including PSNR, SSIM, MS-SSIM, VIF (Sheikh & Bovik, 2016), and VMAF (Li, Aaron, Katsavounidis, Moorthy, & Manohara, 2016) between compressed and original image sequences, at the target bit rates mentioned. For HDR/WCG images, quality metrics include PQ-PSNR-Y, PQ-MS-SSIM-Y (in PQ space), and HDR-VDP (linear space) (Hanhart, Bernardo, Pereira, Pinheiro, & Ebrahimi, 2015).

13.9 Subjective Quality Testing

Subjective quality evaluation of the compressed images must be performed on test images. Testing methodologies include double stimulus impairment scale (DSIS) (ITU-R-BT.500-13, 2012) and absolute category rating with hidden reference (ACR-HR), with a randomized presentation order, as described in ITU-T P.910 (ITU-T-P.910, April 2008).

13.10 Evaluation Tools

To ease the objective assessment of the different proposals, a Docker (Docker Inc., 2019) container and set of Python scripts are provided to automatically perform the objective assessment of a given set of codecs. The Docker container can run on different platforms, including Windows, Ubuntu, and macOS. The source code and installation instructions are available at https://github.com/Netflix/codec_compare.

14

JPSearch

The objective of JPSearch (JPEG, Overview of JPSearch, 2018) is to address interoperability in image search and retrieval systems. For this purpose, JPSearch puts forward an abstract image search and retrieval framework. Interfaces and protocols for data exchange between the components of this architecture are standardized, with minimal restrictions on how these components perform their respective tasks. The use and reuse of metadata and associated metadata schema is thus facilitated. A common query language enables search over distributed repositories. Finally, an interchange format allows users to easily import and export their data and metadata among different applications and devices. In the JPSearch framework, interoperability can be defined in different ways: between self-contained vertical image search systems providing federated search, between layers of an image search and retrieval system so that different modules can be supplied by distinct vendors, or at the metadata level such that different systems may add, update, or query metadata (JPEG, Overview of JPSearch, 2018).

JPSearch is an initiative of the JPEG Committee to address the management and exchange of metadata of images through JPSearch schema and ontology building blocks, a query format, a file format for metadata embedded in image data, and a data interchange format for image repositories. The goal of the standard is to address interoperability in image search and retrieval systems.

The amount of imagery available through various content providers grows at a staggering rate. Kumar, S.N., Bharadwaj, M.V. and Subbarayappa, S., 2021, April. Performance Comparison of Jpeg, Jpeg XT, Jpeg LS, Jpeg 2000, Jpeg XR, HEVC, EVC and VVC for Images. In 2021 6th IEEE International Conference for Convergence in Technology (I2CT) (pp. 1–8). However, the lack of consistency in how repositories are accessed for retrieval or syncing complicates interoperability between devices and systems. For example, metadata could be changed, i.e., when using personal computers or cloud computing infrastructure because the metadata formats may be different or, simply, the import and export schemes are non-compliant.

Contemporary online and offline photo management applications do not use a standardized way to manage and exchange annotations with other systems. Most content providers employ dedicated application programming interfaces (APIs) to access their content.

JPSearch defines the components of a framework that enables interoperability among image repositories and/or their clients, by defining interfaces and protocols for data exchange between devices and systems, while restricting as little as possible how those devices, systems, or components perform their task (JPEG, Overview of JPSearch, 2018).

14.1 JPSearch Parts

JPSearch is a multi-part standard and currently covers six parts as follows:

Part 1.   Global architecture: Provides an overview of the global architecture of the JPSearch framework. In addition, Part 1 provides use cases and examples to understand the use of the JPSearch framework.

Part 2.   Schema and ontology: Describes the registration, identification, and management of schema and ontology.

Part 3.   Query format: Specifies the JPSearch Query Format and JPSearch API for querying still image repositories. Both textual and visual queries are supported.

Part 4.   File format: Supports the creation and maintenance processes by defining a file format for metadata embedded in image data (typically, JPEG and JPEG 2000).

Part 5.   Data interchange format: Enables synchronization by defining a data interchange format between image repositories.

Part 6.   Reference software: Provides reference software that instantiates the functionality defined in the earlier parts.

Each part is described next.

14.1.1 Part 1: Global Architecture

The first part provides a global view and introduction of the JPSearch specification (ISO/IEC-24800-1, 2012) and is constructed such that it integrates smoothly in typical image processing and management architectures, enabling bilateral exchange of information between content producers, consumers, and/or aggregators as shown in Figure 14.1.

Part 1 was published as a technical report (Leong, Chang, Houchin, & Dufaux, 2007) that defines several use cases which should be supported by the framework; some of them are mentioned as follows:

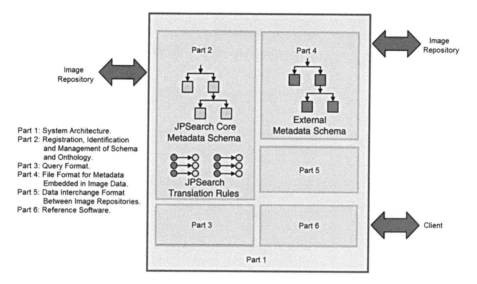

Figure 14.1 JPSearch system global architecture.

Searching images in stock photo collections for usage in magazines: A user wishes to buy a selection of images in order to illustrate a publication to be sold to consumers.

Mobile tourist information: A tourist takes a picture of the landmark on a mobile phone and sends it to a tourist information server, which calls back to the tourist and gives him/her the information.

Finding illegal or unauthorized use of images: The user holds his original content and he wants to find unauthorized variations of his original content using search engines.

Matching images between collections for synchronization: The user ends up with a large collection of images stored on multiple devices and he needs his collection to be synchronized across all platforms.

Image search in the medical domain: Many pathologies have visual symptoms, which are essential for doctors doing diagnosis. The user (a doctor in a hospital clinic) searches for the best matches of a pathology related to symptoms and retrieves case histories (including other images, metadata, text, etc.) to improve his/her diagnosis.

Figure 14.2 shows an example of how JPSearch files are embedded in JPEG or JPEG 2000 file formats, with the potential to have multiple JPSearch files carried by one image file. Each JPSearch file can contain, in its turn, multiple JPSearch Core Metadata schema and registered schema. Each data block is an instance of a JPSearch Core Metadata schema (at least one) or

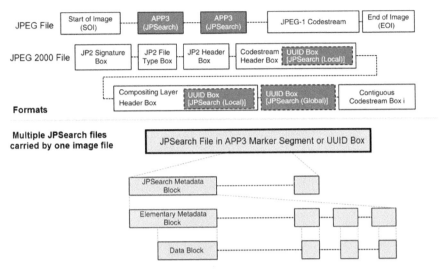

Figure 14.2 JPEG or JPEG 2000 file formats, with the potential to have multiple JPSearch files.

JPSearch registered schema each associated with a specific, potentially different author.

14.1.2 Part 2: Schema and Ontology

Part 2 consists of three main tools to support interoperability between image descriptions based on different metadata schema (ISO/IEC-24800-2:2011, 2011). The JPSearch Core Schema serves as root schema for any image description and it should be understood by any retrieval system. The translation mechanism provides means for mapping descriptive information between various XML-based metadata models, and, finally, the management unit deals with registered metadata models and its translation rules. In order to provide interoperable search functionality among the different repositories or systems, one common metadata schema, JPSearch Core Metadata, is utilized for delivering a user's query in an understandable manner among the systems that process search-related operations using the different metadata schema.

The core metadata provides definitions of essential metadata describing images and a reference in describing third-party metadata with help translation rules. Table 14.1 shows the descriptions of the JPESearch core elements. The JPSearch Translation Rules Declaration Language (JPTRDL) is defined to enable specifying guidelines for translating third party defined metadata schema to core metadata. The management tools register, update, and

Table 14.1 Descriptions of the JPSearch core elements.

Element: Description
Identifier: Identifies an individual image
Modifiers: Names of the people who modified the image
Creators: Name of the person who cr eated the image
Publisher: Name of the person or organization who made the image available
CreationDate: Date of creation
ModifiedDate: Date of modification
Description: Free text description of the image content
RightsDescription: Rights-related information using either free text or another rights description standard
Source: Information about the source of the image
Keyword: List of keywords characterizing the image
Title: Title of the image
CollectionLabel: Name of the image's collection
PreferenceValue: Level of preference on the image given by the metadata provider
Rating: Rating information providing definition and value of the rating based on the controlled term
OriginalImageIdentifier: Identifier of the image from which the given image is originated
GPSPositioning: Location shown in the image
RegionOfInterest: Description of certain regions of the image with additional information including description using external metadata standards as well as content description of person, object, place, and event
Width: Width of the image in number of pixels
Height: Height of the image in number of pixels

retrieve third-party metadata with translation rule to the JPSearch metadata authority. In this case, users can understand metadata instances that are not defined using JPSearch Core Metadata (Yoon *et al.*, 2012). The JPSearch Core Schema is modeled as an XML schema and provides a light-weight and extensible mechanism for annotating image content. The current version supports 20 main elements, chosen according to their coverage by popular metadata formats, including spatial and region of interest-based annotation (Temmermans *et al.*, 2013).

However, the core schema neither is intended to be the unique schema used for annotating images nor is it intended to replace existing well-established metadata schema such as MPEG-7 (MPEG, 1988) or (DCIM,

1995). The JPSearch Core Metadata contains a set of minimal core terms that can be extended in two different ways: by extending the Core Schema through External Description elements and by directly using a different metadata schema in the JPSearch-based interactions (annotation, querying, etc.) and defining and registering the proper translations to the core schema with the JPSearch Translation Rules Declaration Language. Figure 14.3 shows an example of an image annotation.

The code of Figure 14.3 contains administrative information, like Creators or CreationDate, and basic descriptive annotations about the content, such as Description and Keyword. It also provides technical information about the size (Width and Height). Finally, it shows how specific salient and interesting regions within an image are denoted, using the RegionOfInterest element. This element supports the specification of events, persons, or places (JPSearch-RA, 2011). Figure 14.4 shows an example of a code that describes a fictional image.

The Translation Rules Declaration Language (TRDL) provides means for translating descriptive information within JPSearch queries to and from the JPSearch Core Schema into equivalent information of a respective target schema. TRDL focuses on mappings at the structural and syntactic level for transforming the descriptive content of queries. In other words, TRDL describes corresponding relationship between JPSearch Core Metadata Schema and a registered metadata schema such as one-to-one, one-to-many, or many-to-one relationship in the context of XML element level. To support the relationship description, JPSearch defines three types: OneToOneFieldTranslationType, ManyToOneFieldTranslationType, and OneToMany-FieldTranslationType. The position of a particular XML element is described by using XPath expression. Moreover, JPSearch metadata translation rule system provides mechanism for complex selection of elements in corresponding relationship definition by using a type FilteredSourceFieldType that supports regular expression, variable usage, position-based binding, and list-based binding (Yoon *et al.*, 2012).

The objective of Part 2 is to correctly understand and handle XML-based metadata expressed using different proprietary schema (ISO/IEC-24800-4:2010, 2010). The JPSearch Core Metadata schema serves as the underlying foundation for interoperable search and retrieval operations across several image repositories. It is used in conjunction with the JPSearch Query Format, as defined in Part 3, to formulate requests to JPSearch compliant systems.

```
<?xml version="1.0" encoding="UTF-8"?>
<JPCore:ImageDescription>
 <JPCore:Creators>
 <JPCore:GivenName>Mario</JPCore:GivenName>
 <JPCore:FamilyName>Doeller</JPCore:FamilyName>
 </JPCore:Creators>
 <JPCore:CreationDate>2001-12-17T09:30:47.0Z</JPCore:CreationDate>
 <JPCore:Description>Soccer Game</JPCore:Description>
 <JPCore:Keyword>Fans</JPCore:Keyword>
 <JPCore:Keyword>Soccer Game</JPCore:Keyword>
 <JPCore:GPSPositioning latitude="-90.0" longitude="-180.0" altitude="3.1415"/>
 <JPCore:RegionOfInterest>
 <JPCore:RegionLocator>
  <JPCore:Region dim="2">0 0 20 20</JPCore:Region>
 </JPCore:RegionLocator>
 <JPCore:Description>Spectators are crying</JPCore:Description>
 <JPCore:ContentDescription>
 <JPCore:Person>
 <JPCore:Name><JPCore:FamilyName>Spectators
 </JPCore:FamilyName></JPCore:Name>
 </JPCore:Person>
 <JPCore:Place>
 <JPCore:Name>Stadium</JPCore:Name>
 </JPCore:Place>
 <JPCore:Event>
 <JPCore:Label>urn:test:cs:cheering</JPCore:Label>
 </JPCore:Event>
 </JPCore:ContentDescription>
 </JPCore:RegionOfInterest>
 <JPCore:Width>255</JPCore:Width>
 <JPCore:Height>255</JPCore:Height>
 </JPCore:ImageDescription>
```

Figure 14.3   Example of image annotation.

```
<?xml version="1.0" encoding="iso-8859-1"?>

<ImageDescription xmlns="urn:jpeg:jpsearch:schema:coremetadata:2009"
xmlns:xsi="http://www.w3.org/2001/XMLSchema-instance"
xsi:schemaLocation="urn:jpeg:jpsearch:schema:coremetadata:2009 24800-2-core.xsd">

<Identifier>urn:unique:identifier:1:2:3</Identifier>

<Creators>

<GivenName>John</GivenName>

<FamilyName>Smith</FamilyName>

</Creators>

<CreationDate>2011-12-17T09:30:47.0Z</CreationDate>

<ModifiedDate>2011-12-17T09:30:47.0Z</ModifiedDate>

<Description>Sample description</Description>

<Keyword>Sardinia</Keyword>

<Keyword>Italy</Keyword>

<Keyword>50th JPEG meeting</Keyword>

<Title>Example Instance document of the JPSearch core schema</Title>

<GPSPositioning latitude="34" longitude="34" altitude="10"/>

<RegionOfInterest>

<RegionLocator>

<Region dim="2"> 0 0 100 100</Region>

</RegionLocator>

<Description>A short description about the selected region

</Description>

<Keyword>plenary meeting</Keyword>

</RegionOfInterest>

<Width>640</Width>

<Height>480</Height>

</ImageDescription>
```

Figure 14.4   Example of the description of a fictional image (JPSearch-RA, 2011).

14.1.3 Part 3: Query Format

Part 3 specifies message format in XML schema to be sent and received between information requestors (clients) and information providers (database servers) (ISO/IEC-24800-3:2010, 2015). It is an XML-based query language that defines the syntax of queries, exchanged between client applications and repositories. The JPSearch Query Format (JPQF) facilitates and unifies access to search functionalities in distributed digital image repositories and it is defined as a subset of the MPEG7 Query Format (MPQF), restricted to the image domain. The query format consists of three parts: the input query format, the output result format, and the query management tools. The input query format provides means for describing query requests from a client to an image repository. The output query format specifies a message container for retrieval responses. Finally, the query management tools arrange all organizational aspects of sending a query to a repository. This includes functionalities such as service discovery, service aggregation, and service capability description. An input query can be composed of three different parts: a declaration, an output description, and a query condition. The declaration contains references to resources, which can be used within the output description or query condition. Such a reference can, for example, point to an image file or a metadata description file. The output description allows to define the structure as well as the content of the expected result set. Finally, the query condition or query management part provides tools related to the organizational aspects of exchanging queries. This includes functionalities such as service discovery, service aggregation, and service capability description.

14.1.4 Part 4: File Format

Part 4 specifies file formats based on JPEG file format and JPEG 2000 file format to carry metadata with the images (ISO/IEC-24800-4:2010, 2010). These file formats allow multiple metadata to be embedded inside the JPEG file or JPEG 2000 file. Social tagging is supported by enabling multiple occurrences of metadata inside an image file. These heterogeneous metadata can have essential interoperability through the translation rules defined in Part 2. The processed query result is formed into a query output as defined in Part 3 and returned to the user.

There are two types of JPSearch file format; one is fully compatible to JPG (JPEG) file format and the other is to J2K (JPEG 2000) file format. Both formats have a common format independent blocks named as JPSearch metadata block (JPS-MB) to carry metadata instances. JPS-MB is stored in JPEG file format using application marker segments, APP3, as shown in Figure 14.5.

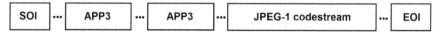

Figure 14.5 Structure of JPEG compatible JPSearch file format.

A JPS-MB is a binary bit stream and designed to carry multiple instances of elementary metadata (JPS-EM). When the size exceeds the limitation, several sets of JPS-MB, each of which has different JPS-EM, shall be employed. An APP3 segment can have only one JPS-MB inside. The JPS-EM consists of identifier of the used schema, several descriptive information of metadata creation process (creation/update date/time, author information and confidence of the description), and metadata itself. If another instance using different schema is required, the user can easily add another JPS-EM into a JPS-MB (Yoon *et al.*, 2012).

JPS-MB is stored in JPEG 2000 file format using Universally Unique Identifier boxes (UUID) as shown in Figure 14.6.

Since the file format allows multiple codestreams, after the JP2 Header Box, two different types of UUID boxes can be instantiated. The first UUID box at the top level is used to describe globally applicable metadata and those in codestream-layer headers are used to describe the corresponding local codestream. If local information is different from the global one, the local description overwrites the global one.

In general, JPSearch Part 4 provides metadata carriage capability by embedding the metadata in image data itself so that the user never worries about the loss of achievements when he/she wants to move the content from a specific service provider to another. The benefits of the standardized format are: (a) the carriage of the metadata and the content, (b) compatible files with existing file formats, (c) the carriage of any type of metadata instances that

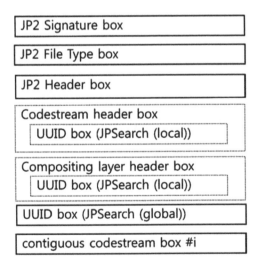

Figure 14.6 Structure of JPEG 2000 compatible JPSearch file format.

can be understood by other users, and (d) multiple instantiations of metadata and different schema to support social tagging for the use of images in a community (Yoon *et al.*, 2012).

14.1.5 Part 5: Data Interchange Format

JPSearch (ISO/IEC-24800-5:2011, 2011) data interchange format enables a number of functionalities to improve the portability metadata associated with an image or a collection of images (Yoon *et al.*, 2012).

Only JPEG-conforming (ISO/IEC-10918, 1994), JPEG 2000-conforming (ISO/IEC-15444, 2004), and JPSearch-conforming codestreams may be used as internal or external resources in the JPSearch data interchange file format.

JPSearch encompasses either internal or external resources, namely encoded image bit streams, along with collection-level and item-level metadata, namely entity of descriptive data of image data.

The format encapsulates all the collection-level and item-level metadata into a valid XML descriptor in a single text-based format. The collection-level and item-level metadata should follow the interchange format schema.

JPSearch enables the synchronization of repositories in order to facilitate simple and fully interoperable exchanges across different devices and platforms.

The file format enables the easy and reliable transfer of data between different hardware and software systems. It should support functions such as:

- exchange of data between JPSearch repositories on different devices and platforms;
- consolidation of metadata generated on different systems;
- transfer of data to a newer and better system;
- consolidation of selected data to a centralized, repository;
- archive of data in a format, which will survive current products.

The XML metadata descriptions for JPSearch collection metadata schema and JPSearch XML metadata interchange format schema are explained as follows:

The type `ImageDataType` allows linking between metadata and each corresponding internal or externa resource when used within the binary file format or just external resources when used within the XML metadata interchange format. `InlineMedia` and `MediaUri` of `ImageDataType` are instances of an image codestream used to store internal resources and to identify the location of external resources (where the image file is located), respectively.

As a special case, a file can have all images defined as external resources. As a result, the file only contains metadata (collection-level and item-level).

The `Collections` element serves as the root element of the collection metadata schema. The root element shall be used as the topmost element when collections metadata appears in an independent way. The `CollectionsType` type is the root element of the collection metadata part of the metadata interchange format schema and allows expressing metadata related to one image collection.

The `ImageRepository` element serves as the root element of the XML metadata interchange format schema. It corresponds to the movie box as defined in the ISO base media file format. The `ImageRepositoryType` type allows expressing metadata related to images and image collections. It is composed of zero-to-many `CollectionsMetadata` elements and one-to-many Image elements. It corresponds to the movie box type as defined in the ISO base media file format (ISO/IEC-15444-12).

The `CollectionsMetadataType` type allows expressing metadata related to image collections. It is composed of multiple `Collection` elements from the `CollectionType` defined in the JPSearch collection metadata schema.

The `ImageType` type allows expressing metadata related to one image.

The `ImageDataType` type allows linking between metadata and each corresponding internal or external resource when used within the binary file format or just external resources when used within the XML metadata interchange format.

The `ImageMetadataType` type allows expressing metadata related to one image (item level). It includes descriptive elements from the JPSearch Core Schema in JPSearch Part 2 (ISO/IEC-24800-2:2011, 2011) but also offers the possibility to include metadata formalized according to external or user defined schemas (e.g., MPEG-7) (Yoon *et al.*, 2012).

14.1.6 Part 6: Reference Software

The reference software generates conformant JPSearch metadata and image files. The ISO/IEC 24800-6:2012 describes reference software for the normative clauses as well as utility software demonstrating the usage scenarios of ISO/IEC 24800-2 to ISO/IEC 24800-5. The reference software helps implementers interpreting the ISO/IEC 24800 specifications and to enable them determining whether their products or systems are conformant to the standard.

The reference software is entirely written in the Java programming language, and it is divided into four different modules, one for each normative

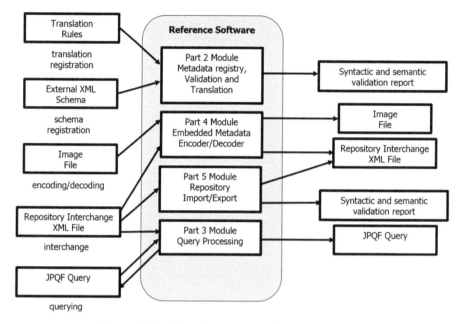

Figure 14.7   JPSearch reference software architecture.

part of the standard (Parts 2, 3, 4, and 5) as shown in Figure 14.7 (Yoon *et al.*, 2012).

Part 2 module registers external metadata schemas, validates metadata instances, and defines cross schema metadata translation. Also, the module performs syntactic and semantic schema/translation validation to determine whether the schema and translations are conformant to the rules in (ISO/IEC-24800-2:2011, 2011).

Part 3 module performs query validation and accepts queries containing Boolean operators, XPath expressions, arithmetic and comparison expressions, sorting, and grouping. The module returns the query evaluation as a JPSearch Query Format (JPQF) message. The module processes JPSearch queries as defined in (ISO/IEC-24800-3:2010, 2015) against a given image repository provided by the user in the form of an (ISO/IEC-24800-5:2011, 2011) file.

Part 4 module accepts either an already annotated image file or a pair (image file, ISO/IEC 24800-5 file) and returns either an ISO/IEC 24800-5 file or an annotated image file, respectively. The module allows extracting/annotating metadata embedded within an image file.

Part 5 module imports/exports metadata from an internal repository. It accepts ISO/IEC 24800-5 files and performs syntactic validation.

15

JPEG Systems

JPEG Systems (JPEG, Overview of JPEG Systems, 2018) intends further to specify system layer extensions for JPEG standards to enable generic support for interactivity protocols, high dynamic range (HDR) tools, privacy and security, metadata, 360° images, augmented reality, and 3D.

As a first result, the principles for the system layer structure of JPEG standards are now defined in Part 1 – "Packaging of information using codestreams and file formats" (ISO/IEC TR 19566-1:2016) and Part 2 – "Transport mechanisms and packaging" (ISO/IEC TR 19566-2:2016). JPEG Systems is also a multi-part specification and currently includes the following six parts.

15.1 JPEG Systems Parts

JPEG Systems, like most other JPEG standards, is a multi-part specification. JPEG Systems currently includes the following parts:

Part 1. Packaging of information using codestreams and file formats: JPEG Systems Part 1 is a technical report describing the packaging of information using codestreams and file formats in legacy formats and give guidelines for future standards.

Part 2. Transport mechanisms and packaging: JPEG Systems Part 2 is a technical report describing the transport mechanisms and packaging in legacy formats and give guidelines for future standards.

Part 3. Feature list and boxtype ID (dropped): JPEG Systems Part 3 was intended to list all features and boxtype IDs in JPEG standards. Due to frequent modifications, this part was dropped and added to the registration authority website.

Part 4. Privacy, security, and IPR features: JPEG Systems Part 4 adds extensions to JPEG standards for privacy, security, and IPR features. An example is the addition of encrypted supplementary images, which can replace scrambled image parts in the base image. It is based on the framework of JPEG Systems Part 5.

Part 5.   JPEG Universal Metadata Box Format (JUMBF): JPEG Systems Part 5 specifies a framework for JPEG standards to add universal metadata allowing future extensions using metadata, supplementary images, or other elements in addition to the base image.

Part 6.   JPEG 360: JPEG Systems Part 6 defines a method to represent 360° images in JPEG standards and to add supplementary metadata and images. It is based on the framework of JPEG Systems Part 5.

16

JBIG

JBIG was prepared by the Joint Bi-Level Image Experts Group (JBIG) of ISO/IEC JTC1/SC29/WG9 and CCITT SGVIII. The JBIG experts group was formed in 1988 to establish a standard for the progressive encoding of bi-level images (ITU-T-T.82/ISO-IEC-11544, 1993) (JPEG, 2019).

JBIG (also called JBIG1) never achieved the expected acceptance even though it provided a 20%–30% reduction in file size over TIFF G4 because of JBIG formats' ability to use arithmetic coding instead of Huffman coding. Arithmetic coding allows for data to be represented by a fraction of a bit. Huffman coding is used in TIFF G4 and it requires whole bits to represent runs in the image (Kohn, 2005). JBIG was mostly used for bitonal image compression in a very limited range of (mostly Japanese) MFP devices and digital copiers.

In contrast, the digital media industry has readily received the JBIG2 standard. Almost from the time of its introduction, JBIG2 was supported for bitonal compression in the JPEG 2000 Part 6 specifications, and as a compression filter in Adobe PDF (VeryDOC Company, 2002). It quickly became the format of choice for a number of document-heavy organizations including legal, media, financial, scanning, and banking firms (CVISION Technologies, 1998).

Compression of this type of image is also addressed by existing standards, for example, MH&MR (ITU-T T.4), MMR (ITU-T T.6), and JBIG1 (T.82| ISO/ IEC 11544). The JBIG2 was prepared for lossy, lossless, and lossy-to-lossless image compression and allow for lossless compression performance better than that of the existing standards and to allow for lossy compression at much higher compression ratios than the lossless ratios of the existing standards, with almost no visible degradation of quality by using pattern matching and substitution techniques in addition to the technologies of the existing standards (Howard, Kossentini, Marins, Forchhammer, & Rucklidge, 1998).

Besides the obvious facsimile application (Howard, Kossentini, Marins, Forchhammer, & Rucklidge, 1998), (Ono, Rucklidge, Arps, & Constantinescu,

2000), (de Queiroz, 1999), (Tompkins & Kossentini, 1999), (Simard, Malvar, Rinker, & Renshaw, 2004), (Ye & Cosman, 2003), JBIG2 is useful for document storage and archiving, coding images on the World Wide Web, wireless data transmission, print spooling, and even teleconferencing (ITU-T-T.88/ISO-IEC-14492). Currently, PDF files having versions 1.4 and above may contain JBIG2-compressed data. Multiple colors can be handled by using an appropriate higher-level standard such as ITU-T Recommendation T.44 (JPEG, 2019).

This chapter describes the Rec. ITU-T T.88 | ISO/IEC 14492, commonly known as JBIG2. Implementers are cautioned that this may not represent the latest information and are therefore strongly urged to consult the TSB patent database at http://www.itu.int/ITU-T/ipr/.

16.1 Subject Matter for JBIG Coding

JBIG2 encodes bi-level documents with one or more pages with some text data of a small size. Text data is arranged in horizontal or vertical rows. The characters in the text part of a page are called *symbol*. A page may also contain gray-scale or color multi-level images (e.g., photographs), also called halftone data, that have been dithered to produce bi-level images.

The periodic bitmap cells in the halftone part of the page are called *patterns* in JBIG2. In addition, a page may contain other data, such as line art and noise. Such non-text, non-halftone data is called *generic* data in JBIG2. Text data and halftone data are treated as special cases.

The encoder divides the content of a page into a text region, containing digitized text, a halftone region, containing digitized halftones, and a generic region, containing the remaining digitized image data, such as line art. In some circumstances, it is better (in image quality or compressed data size) to consider text or halftones as generic data; on the other hand, in some circumstances, it is better to consider generic data using one of the special cases (ITU-T-T.88/ISO-IEC-14492, 2018).

A text region consists of a number of symbols placed at specified locations on a background. The symbols usually correspond to individual text characters. JBIG2 reuses symbols by collecting them into a symbol dictionary to increase its effectiveness. A halftone region consists of a number of patterns placed along a regular grid. The patterns usually correspond to gray-scale values.

16.2 Relationship Between Segments and Documents

Typically, a page is encoded using segments. Segments are numbered sequentially and their structure consists of a segment header, a data header, and the data. The segment header stores the segment reference information (in case of multi-page documents, the page association information). A data header provides information to encode the data in the segment. The data describes an image region or a dictionary or provides other information. The segments are the following:

1. Page information segment that provides information about the page (size and resolution).
2. Symbol dictionary segment that gathers bitmaps referred to in the region segments. A dictionary segment may be associated with one page of the document or with the document as a whole. Also, this segment may refer to earlier dictionary segments. The symbols added to a dictionary segment may be described directly or may be described as refinements of symbols described previously, either in the same dictionary segment or in earlier dictionary segments.
3. Text region segment that defines the appearance of the text and halftone regions by referencing bitmaps from a dictionary and specifies the place where they should appear on the page. A region segment is always associated with one specific page in the document and it may refer to one or more earlier dictionary segments. This reference allows the decoder to identify symbols in a dictionary segment that are present in the image. A region segment also may refer to an earlier region segment. This reference allows to combine the image described by the earlier segment with the current representation of the page.
4. Pattern dictionary segment.
5. Halftone region segment.
6. End-of-page segment that specifies the end of page.

Some segments give information about the structure of the document: start of page, end of page, and so on. Some segments code regions used in turn to produce the decoded image of a certain page. Some segments ("dictionary segments") do neither, but instead define resources that can be used by segments that code regions.

A segment can be associated with some page or not associated with any page. A segment can refer to other, preceding, segments. A segment also includes retention bits for the segment that it refers to and for itself; these indicate when the decoder may discard the data created by decoding a segment.

A segment's header part always begins and ends on a byte boundary. A segment's data part always begins and ends on a byte boundary. Any unused bits in the final byte of a segment must contain 0 and shall not be examined by the decoder.

The segment header part and the segment data part of a segment need not occur contiguously in the bit stream being decoded.

A JBIG2 file may be organized in three ways, sequential, random access, or embed into a non-JBIG2 file (embedded organization). In the sequential organization, each segment's header immediately precedes that segment's data header and data, all in sequential order.

In the random access organization, all the segment headers are collected together at the beginning of the file, followed by the data (including data headers) for all the segments, in the same order. This second organization permits a decoder to determine all segment dependencies without reading the entire file.

In the embedded organization case, a different file format carries JBIG2 segments. The segment header, data header, and data of each segment are stored together, but the embedding file format may store the segments in any order, at any set of locations within its own structure.

16.3 JBIG2 Decoder

Figure 16.1 shows the major decoder components and associated buffers. The memory components are: symbol memory, context memory, pattern memory, and page and auxiliary buffers (ITU-T-T.88/ISO-IEC-14492, 2018). One decoding procedure invokes another decoding procedure (solid lines). For example, the symbol dictionary decoding procedure invokes the generic region decoding procedure to decode the bitmaps for the symbols that it defines.

Dashed lines indicate flow of data. For example, the text region decoding procedure reads symbols from the symbol memory and sends them to the page buffer or an auxiliary buffer. The page and auxiliary buffers recover the final decoded page images.

It is estimated that a full-featured decoder may need two full-page buffers plus about the same amount of dictionary memory, plus about 100 kilobytes of arithmetic coding context memory, to decode most bit streams (ITU-T-T.88/ISO-IEC-14492, 2018).

A buffer is a representation of a bitmap and is intended to hold a large amount of data (typically the size of a page to contain the description of a region or of an entire page). Even if the buffer describes only a region, it has information associated with it that specifies its placement on the page. Decoding a region segment modifies the contents of a buffer.

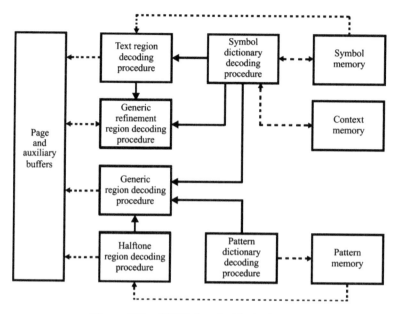

Figure 16.1   JBIG2 decoder block diagram.

All buffers are auxiliary buffers except the page buffer. The page buffer is used by the decoder to accumulate region data until the page has been completely decoded; then the data can be sent to an output device or file. The decoder may output an incomplete page buffer, either as part of progressive transmission or in response to user input. Such is not specified in the standard.

A symbol dictionary consists of an indexed set of bitmaps. The bitmaps in a dictionary are typically small, approximately the size of text characters. Unlike a buffer, a bitmap in a dictionary does not have page location information associated with it.

Table 16.1 summarizes the types of data being decoded, which decoding procedure is responsible for decoding them, and what the final representations of the decoded data are.

Decoding a segment involves invocation of one or more decoding procedures. The decoding procedures to be invoked are determined by the segment type.

The result of decoding a region segment is a bitmap stored in a buffer, possibly the page buffer. Decoding a region segment may fill a new buffer or may modify an existing buffer. In typical applications, placing the data into a buffer involves changing pixels from the background color to the

Table 16.1 Entities in the decoding process (ITU-T-T.88/ISO-IEC-14492, 2018).

Concept	JBIG2 bit-stream entity	JBIG2 decoding entity	Physical representation
Document	JBIG2 file	JBIG2 decoder	Output medium or device
Page	Collection of segments	Implicit in control decoding procedure	Page buffer
Region	Region segment	Region decoding procedure	Page buffer or auxiliary buffer
Dictionary	Dictionary segment	Dictionary decoding procedure	List of symbols
Character	Field within a symbol dictionary segment	Symbol dictionary decoding procedure	Symbol bitmap
Gray-scale value	Field within a halftone dictionary segment	Pattern dictionary decoding procedure	Pattern

foreground color, but this Recommendation|International Standard specifies other permissible ways of changing a buffer's pixels.

A typical page will be described by one or more immediate region segments, each one resulting in modification of the page buffer.

Just as it is possible to specify a new symbol in a dictionary by refining a previously specified symbol, it is also possible to specify a new buffer by refining an existing buffer. However, a region may be refined only by the generic refinement decoding procedure. Such a refinement does not make use of the internal structure of the region in the buffer being refined. After a buffer has been refined, the original buffer is no longer available.

The result of decoding a dictionary segment is a new dictionary. The symbols in the dictionary may later be placed into a buffer by the text region decoding procedure.

16.3.1 Decoding Procedure

The generic region decoding procedure fills or modifies a buffer directly, pixel-by-pixel if arithmetic coding is being used, or by runs of foreground and background pixels if MMR and Huffman coding are being used. In the arithmetic coding case, the prediction context contains only pixels determined by data already decoded within the current segment.

The generic refinement region decoding procedure modifies a buffer pixel-by-pixel using arithmetic coding. The prediction context uses pixels determined by data already decoded within the current segment as well as pixels already present either in the page buffer or in an auxiliary buffer.

The text region decoding procedure takes symbols from one or more symbol dictionaries and places them in a buffer. This procedure is invoked during the decoding of a text region segment. The text region segment contains the position and index information for each symbol to be placed in the buffer; the bitmaps of the symbols are taken from the symbol dictionaries.

The symbol dictionary decoding procedure creates a symbol dictionary, that is, an indexed set of symbol bitmaps. A bitmap in the dictionary may be coded directly; it may be coded as a refinement of a symbol already in a dictionary or it may be coded as an aggregation of two or more symbols already in dictionaries. This decoding procedure is invoked during the decoding of a symbol dictionary segment.

The halftone region decoding procedure takes patterns from a pattern dictionary and places them in a buffer. This procedure is invoked during the decoding of a halftone region segment. The halftone region segment contains the position information for all the patterns to be placed in the buffer as well as index information for the patterns themselves. The patterns, the fixed-size bitmaps of the halftone, are taken from the halftone dictionaries.

The pattern dictionary decoding procedure creates a dictionary, that is, an indexed set of fixed-size bitmaps (patterns). The bitmaps in the dictionary are coded directly and jointly. This decoding procedure is invoked during the decoding of a pattern dictionary segment.

The control decoding procedure decodes segment headers, which include segment type information. The segment type determines which decoding procedure must be invoked to decode the segment. The segment type also determines where the decoded output from the segment will be placed. The segment reference information, also present in the segment header and decoded by the control decoding procedure, determines which other segments must be used to decode the current segment. The control decoding procedure affects everything shown in Figure 16.1 and so is not shown as a separate block.

16.4 Lossy Coding

Lossy symbol coding provides a natural way of doing lossy coding of text regions (ITU-T-T.88/ISO-IEC-14492, 2018). The idea is to allow small differences between the original symbol bitmap and the one indexed in the symbol dictionary. Compression gain is affected by not having to code a large dictionary and by having a cheap symbol index coding as a consequence of the smaller dictionary. It is up to the encoder to decide when two bitmaps are essentially the same or essentially different (Ascher & Nagy, 1974).

16.4.1 Symbol Coding

The hazard of lossy symbol coding is to have *substitution errors*, that is, to have the encoder replace a bitmap corresponding to one character by a bitmap depicting a different character so that a human reader misreads the character.

The risk of substitution errors can be reduced by using intricate measures of difference between bitmaps and/or by making sure that the critical pixels of the indexed bitmap are correct. One way to control this, described in Howard (1996), is to index the possibly wrong symbol and then to apply refinement coding to that symbol bitmap. The idea is to encode the basic character shape at little cost and then correct pixels that the encoder believes alter the meaning of the character.

The process of beneficially introducing loss in textual regions may also take simpler forms such as removing flyspecks from documents or regularizing edges of letters. Most likely, such changes will lower the code length of the region without affecting the general appearance of the region – possibly even improving the appearance.

A number of examples of performing this sort of lossy symbol coding with JBIG2 can be found in Howard, Kossentini, Marins, Forchhammer, and Rucklidge (1998).

16.4.2 Generic Coding

To effect, near-lossless coding using generic coding (Howard, Kossentini, Marins, Forchhammer, & Rucklidge, 1998), the encoder applies a pre-process to an original image and encodes the changed image losslessly. The difficulties are to ensure that the changes result in a lower code length and that the quality of the changed image does not suffer badly from the changes. Two possible pre-processes are given in Marints and Forchhammer (1997). These pre-processes flip pixels that, when flipped, significantly lower the total code length of the region but can be flipped without seriously impairing the visual quality. The pre-processes provide for effective near-lossless coding of periodic halftones and for a moderate gain in compression for other data types. The pre-processes are not well-suited for error diffused images and images dithered with blue noise as perceptually lossless compression will not be achieved at a significantly lower rate than the lossless rate.

16.4.3 Halftone Coding

Halftone coding (Howard, Kossentini, Marins, Forchhammer, & Rucklidge, 1998) is the natural way to obtain very high compression for *periodic* halftones, such as clustered-dot ordered dithered images. In contrast to lossy generic coding as described above, halftone coding does not intend to preserve the original bitmap, although this is possible in special cases. Loss can also be introduced for additional compression by not putting all the patterns of the original image into the dictionary, thereby reducing both the number of halftone patterns and the number of bits required to specify which pattern is used in which location.

For lossy coding of error diffused images and images dithered with blue noise, it is advisable to use halftone coding with a small grid size. A reconstructed image will lack fine details and may display blockiness but will be clearly recognizable. The blockiness may be reduced on the decoder side in a post-process; for instance, by using other reconstruction patterns than those that appear in the dictionary. Error diffused images may also be coded losslessly or with controlled loss as described above, using generic coding.

16.5 Summary

JBIG2 is an international standard that provides for lossy compression of bi-level images and allows both quality-progressive coding through refinement stages, with the progression going from lower to higher (or lossless) quality, and content-progressive coding, successively adding different types of image data (for example, first text, then halftones).

JBIG2 encoder decomposes the input bi-level image into several regions or image segments (usually based on content), and each of the image segments is separately coded using a different coding method. Such content-based decomposition is very desirable in interactive multimedia applications.

Low-end facsimile requires high coding speed and low complexity, even at the cost of some loss of compression, while wireless transmission needs maximum compression to make the fullest use of its narrow channel. In recognition of the variety of application needs, JBIG2 does not have a baseline implementation.

JBIG2 standard explicitly defines the requirements of a compliant bit stream and implicitly defines a decoder behavior. The standard does not explicitly define a standard encoder but instead is flexible enough to allow sophisticated encoder design. In fact, encoder design is a major differentiator among competing JBIG2 implementations.

ANNEX A. PROVISION

PROVISION – PeRceptually Optimized VIdeo compresSION

Funded by the European Commission's Seventh Framework Programme
Duration: September 2013–August 2017

The PROVISION ITN targeted the important area of perceptually oriented video coding and particularly focused on how to deliver higher volumes of more immersive content over congested band-limited channels. PROVISION delivered world-leading technical innovation through a unique highly integrated training program, producing skilled staff, efficient and validated solutions, and impact through standardization. Only through interdisciplinary collaboration across academia and industry could the required transformational research can be carried out. The projects objectives were to:
- provide the highest levels of technical and transferable skills training to the researcher team;
- conduct training in the context of an international research collaboration between some of the leading universities, research institutes, companies, and end-user enterprises in Europe;
- complement the PROVISION consortium with international experts as associated partners, exposing the young scientists to cultural and organizational variations in research;
- enhance the PROVISION consortium by internationally respected visiting scientists from the public and the private sectors to guarantee optimal conditions for all its researchers (ESR and ER). In the project, five full partners were involved: Fraunhofer HHI (DE), University of Bristol (UK), BBC (UK), RWTH (DE), and University of Nantes (FR). Also, six associated partners – YouTube (US), Netflix (US), Technicolor (FR), VITEC (FR), Queen Marry University (UK), and TU Berlin (DE) – participated in PROVISION, which offered secondments and technical guidance to the project fellows.

To achieve the project goals and objectives, 18 fellows participated in the project. The fellows achieved significant technological progress within their research teams as well as during the individual secondments. Here, scientific advancement in the area of perceptual video coding was achieved through the fellows' individual work, which was structured along five scientific work packages.

The technological highlights in each work package included:
- identification of shortcoming in current video codecs and measurement tool contribution to international standardization in WP1;
- optical-flow-based motion modeling for irregular textures and 3D structural reconstruction for prediction in WP2;
- three-cluster texture model with adapted coding, new content synthesis methods through template matching, and subsampling in WP3;
- visual attention models for 2D and 360° video, development of perceptually optimized intra-coding, new visual disruption quality metric, and non-reference VQA models in WP4;
- successful core validation through oral/poster presentations in international events, demonstrations, subjective assessment, feedback by international experts, and joint topic-based working groups in WP5.

The novelty of the scientific work has been proven through a number of scientific publications, including more than 50 papers at high-level international conferences, such as the ICIP, ICME, ICASSP, MMSP, PCS, EUSIPCO, QoMEX, and international journals, such as *Electronic Imaging*. Several publications were honored with Best Paper Awards. Furthermore, the fellows have provided more than 15 contributions to standard and are co-inventors of 4 patents. In addition to these activities, a novel type of activity, Grand Challenges, also took place.

Here, PROVISION submitted two proposals in 2017 and PROVISION was offered to organize both. Each fellow significantly contributed to the scientific community in image processing and video coding, with the most advanced research topic within PROVISION. In detail, this included the creation and definition of test sets, frameworks, and core experiments, research on rate–distortion curve prediction from uncompressed content and improved reference picture list sorting and signaling.

Furthermore, methods for fast template matching for intra-prediction, multiple transform residual coding and variable bit-rate rate control provided higher compression efficiency in comparison to previous state-of-the-art video codecs. Special emphasis on perception in video coding was given through the research topics on low-complexity spatio-temporal adaptation for video compression, video texture analysis and classification, perceptual RDO augmentation and just-noticeable-difference optimization, and intra-prediction. Detailed work on optimization for motion information in video coding has been done through fellows by fine detail reconstruction for B pictures, motion-based texture synthesis for 2D video content, as well as for 3D content by motion compensation using estimated 3D models. One main topic was the inclusion of perceptual metrics and attention modeling in video coding.

Here, research topics on short-term perceptual fidelity metric for dynamic texture coding, perceptual modeling of temporal structural artifacts, image feature characterization of light-field content, and video chain optimization using content characteristics were carried out. In addition to video coding, the last work item on video chain optimization also targeted transmission optimization. Each of these individual fellow's project advanced the previous state of the art. This could be achieved through PROVISION training activities as a central component in the career development of its researchers, enabling them to enhance existing skills and develop new skills, both related to technical and complementary topics. This included training on different video coding architectures, challenges of commercializing video technology, overview, perspectives, and principles in standardization, as well as on vitae career development, pitching ideas and presentation skills, video making sessions, technical project management, patents and IPR in video compression, writing technical research papers and PhD planning and organization, start-ups and business plans, CV production, and interview techniques.

All fellows gained experience in presenting their work during work package updates and engaged in discussions on progress and planning. Fellows finally gained significant presentation experience through the PROVISION Public Workshop (Berlin) and Summer School on Video Compression and Processing (SVCP) (Kerkrade).

Through their scientific achievements and presentations to the scientific community at respective conferences and standardization meetings, as well as the wider audience at more general events and exhibitions, the fellows highly successfully promoted PROVISION technology beyond the international community in image processing. After the end of their PROVISION engagement, all fellows successfully secured their future technical careers with continuing employments at the full partners' premises, at new employments at technical companies in the area of video compression, or at an own start-up.

Contact

Dr.-Ing. Karsten Müller (Müller, 2019)
Head of Image & Video Understanding Group
Phone +49 30 31002-225
Email: karsten.mueller@hhi.fraunhofer.de

Antje Nestler
Public Affairs & Project Management CINIQ/Smart Data Forum
Phone +49 30 31002-412
Email: antje.nestler@hhi.fraunhofer.de

Further Information

Image & Video Coding Group (IVC) (Fraunhofer HHI, 2019)
Projects of Image & Video Coding Group (Fraunhofer HHI, 2019)

References

Blender Foundation. (2019, 25 July). *Sintel.* Retrieved Jan 25, 2019, from DURIAN Blender open movie project: https://durian.blender.org/sharing/

Fraunhofer HHI. (2019, July 07). *Image & Video Coding.* Retrieved from Fraunhofer Institute for Telecommunications, Heinrich Hertz Institute, HHI: https://www.hhi.fraunhofer.de/en/departments/vca/research-groups/image-video-coding.html

Fraunhofer HHI. (2019, July 18). *Projects of Image & Video Coding Group .* Retrieved from Fraunhofer Institute for Telecommunications, Heinrich Hertz Institute, HHI: https://www.hhi.fraunhofer.de/en/departments/vca/research-groups/image-video-coding/projects.html

ISO/IEC JTC1/SC29/WG 1 M81049. (2018). *JPEG Pleno - Overview of Point Cloud.* Vancouver, Canada.

ISO/IEC JTC1/SC29/WG11, & MPEG2019/N18190. (2019). *V-PCC Codec description (TMC2release v5.0).*

Rec. ITU-T, T. (2016, Aug.). Information technology – JPEG XR image coding system – Image coding specification. International Telecommunication Union.

Adams, M. D. (2013, July 21). Coding of Still Pictures - JBIG - JPEG. *The JPEG-2000 still image compression standard*, 1–16. Victoria, BC, Canada: ISO/IEC JTC 1/SC 29/WG 1 (ITU-T SG 16).

Ahmed, A., Shahid, M., & Rao, K. (1974). Discrete cosine transform. *IEEE Transactions on Computers, C-23*(1), 90–93.

Alexiou, E., & Ebrahimi, T. (2018). Point cloud quality assessment metric based on angular similarity. *IEEE International Conference on Multimedia and Expo (ICME)*, (pp. 1–6).

Arai, Y., Agui, T., & Nakajima, M. (1988, Nov.). A fast DCT-SQ scheme for images. *Transactions of the IEICE, E71*(11), 1095–1097.

ARRI. (2019, July 25). *ARRI.* Retrieved from ARRI: https://www.arri.com/en/

Artusi, A., Mantiuk, R. K., Richter, T., Hanhart, P., Korshunov, P., Agostinelli, M., . . . Ebrahimi, T. (2015, Dec.). Overview and evaluation of the JPEG

XT HDR image compression standard. *Journal of Real-Time Image Processing, 16*(2), 413–428.

Artusi, A., Mantiuk, R. K., Richter, T., Korshunov, P., Hanhart, P., Ebrahimi, T., & Agostinelli, M. (2016, March). JPEG XT: A compression standard for HDR and WCG images [Standards in a Nutshell]. *IEEE Signal Processing Magazine, 33*(2), pp. 118 - 124.

Ascher, R., & Nagy, G. (1974, nOV). A means for achieving a high degree of compaction on scan-digitized printed text. *IEEE Transactions on Computers, C-23*, 1174–1179.

Belyaev, E., Gilmutdinov, M., & Turlikov, A. (2006). Binary arithmetic coding system with adaptive probability estimation by "virtual sliding window". *IEEE International Symposium on Consumer Electronics*, (pp. 1–5). St. Petersburg.

Bi, G., & Zeng, Y. (2003). *Transforms and fast algorithms for signal analysis and representations.* Springer Science.

Bider, D., Ahar, A., Bettens, S., Birnbaum, T., Symeonidou, A., Ottevaere, H., . . . Schelkens, P. (2019, Feb.). Signal processing challenges for digital holographic video display systems. *Signal Processing: Image Communication, 70*(1), 114–130.

Bing, B. (2015). *Next-Generation Video Coding and Streaming.* Wiley.

Biswas, R., Malreddy, S., & Banerjee, S. (2018). A High-Precision Low-Area Unified Architecture for Lossy and Lossless 3D Multi-Level Discrete Wavelet Transform. *IEEE Transactions on Circuits and Systems for Video Technology, 28*(9), 2386–2396.

Bjontegaard, G. (2001). *Calculation of average PSNR differences between RD-curves, ITU-VCEG.*

Blender fundation. (2019, July 25). *Tears of Steel.* Retrieved Jan. 25, 2019, from Project mango: https://mango.blender.org/sharing/

Blinder, D., & et al. (2019). Signal processing challenges for digital holographic video display. *Signal Proces. Image Commun., 70*, 114–130.

Britanak, V., & Rao, K. (2007). *Discrete cosine and sine transform.* Oxford: Academic Press.

Britanak, V., Yip, P., & Rao, K. (2006). *Discrete Cosine and Sine Transforms: General Properties, Fast Algorithms and Integer Approximations.* Oxford: Academic Press.

Bruylants, T., Munteanu, A., Alecu, A., Decklerk, R., & Schelkens, P. (2007). Volumetric image compression with JPEG2000. *Proc. of SPIE* (pp. 1 - 2). San Jose, California: SPIE.

Cabral, B., & Kandrot, E. (2015, Aug. 06). *The technology behind preview photos.* Retrieved 06 15, 2019, from Facebook Code: https://code.fb.com/android/the-technology-behind-preview-photos/

Chen, W.-H., Smith, C., & Fralick, S. (1977, Sept.). A fast computational algorithm for the discrete cosine transform. *IEEE Transactions on Communications, 25*(9), 1004–1009.

Chengjie, T., Srinivasan, S., Sullivan, G., Regunathan, S., & Malvar, H. S. (2008). Low-complexity hierarchical lapped transform for lossy-tolossless image coding in JPEG XR/HD Photo. *Proc. of SPIE Applications of Digital Image Processing . 7073*, pp. 70730C-1 - 70730C-12. San Diego, CA: SPIE.

Chitprasert, B., & Rao, K. (1990). Discrete cosine transform filtering. *In Proc. of Intl. Conf. on Acoustics, Speech, and Signal Processing, 3*, pp. 1281–1284.

Chitprasert, B., & Rao, K. (1990, July). Human visual weighted progressive image transmission. *IEEE Transactions on Communications, 38*(7), 1040–1044.

Christopoulos, C., Askelof, J., & Larsson, M. (2000). Efficient Encoding and Reconstruction of Regions of Interest in JPEG2000. *10th European Signal Processing Conference* (pp. 1–4). Tampere, Finland: IEEE.

CloudCompare. (2019). *3D point cloud and mesh processing software Open Source Project*. Retrieved Dec. 22, 2020, from CloudCompare: https://www.danielgm.net/cc/

CVISION Technologies. (1998). *JBIG2 and JBIG compared*. Retrieved from CVISION Smarter Document Capture: http://www.cvisiontech.com/resources/jbig2-compression-primer/jbig2-and-jbig-compared.html

Daly, S. (1987). *Subroutine for the generation of a two dimensional human visual contrast sensitivity function*. Rochester, NY: Eastman Kodak.

Das, A., Hazra, A., & Banerjee, S. (2010, Feb.). An Efficient Architecture for 3-D Discrete Wavelet Transform. *IEEE Transactions on Circuits and Systems for Video Technology, 20*(2), 286–296.

DCIM. (1995). *Dublin Core Metadata Initiative*. Retrieved Sept. 2019, from Dublin Core: https://www.dublincore.org/themes/development/

de Queiroz, R. (1999). Compression of compound documents. *IEEE International conference on image processing (ICIP)* (pp. 209 - 213). Kobe, Japan: IEEE.

Defaux, F., Le Callet, F., Mantiuk, R. K., & Mrak, M. (2016). *High dynamic range video from acquisition to display and applications*. Kidlington , Oxford, U.K.: Elsevier.

d'Eon, E., Harrison, B., Myers, T., & et al. (2020). *8i Voxelized Full Bodies - A Voxelized Point Cloud Dataset*. ISO/IEC JTC1/SC29 Joint WG11/WG1 (MPEG/JPEG) input document WG11M40059/WG1M74006, Geneva, Italy.

Digital-Imaging-Group. (1997). *FlashPix format specification Ver. 1.0.0.*

Ding, J., Wei, W., & Chen, H. (2011). Context-based adaptive zigzag scanning for image coding. *In Proc. of 2011 Visual Communications and Image Processing (VCIP*, (pp. 1–4). Tainan.

Do, M., M-Maillet, M., & Vetterli, M. (2012, Feb.). On the Bandwidth of the Plenoptic Function. *IEEE Transations on Image Processing*, 708–717.

Docker Inc. (2019). *About Docker CE*. Retrieved from Docker: https://docs.docker.com/install/

Domański, M., Klimaszewski, K., Kurc, M., & et al. (2014). Super-multi-view light-field light field sequence from Poznan University of Technology. *ISO/IEC JTC1/SC29/WG11/M35071*. Poland, Warsaw.

Doufaux, F., Sullivan, G. J., & Ebrahimi, T. (2009, Nov.). The JPEG XR image coding standard. *IEEE Signal Processing Magazine, 26*(1), pp. 195–204.

Drago, F., Myszkowski, K., Annen, T., & Chiba, N. (2003). Adaptive logarithmic mapping for displaying high contrast scenes. In P. Brunet, & D. W. Fellner, *Computer Graphics Forum* (Vol. 22, pp. 419–426). Granada, Spain: Blackwell.

Dufaux, F., Sullivan, G. J., & Ebrahimi, T. (2009). The JPEG XR image coding standard. *IEEE Signal Processing Magazine*, 195 - 204.

Ebrahimi, T. (2019, April 14). *SlideShare*. Retrieved from The next generation JPEG standards: www.slideshare.net/touradj_ebrahimi/the-next-generation-jpeg-standards

EIZO-Corporation. (2018, 12 12). *ColorEdge CG318-4K*. Retrieved from EIZO: https://www.eizoglobal.com/products/coloredge/cg318-4k/index.html

Faria, L., Fonseca, L., & Costa, M. (2012, Oct.). Performance evaluation of data compression systems applied to satellite imagery. *Journal of Electrical and Computer Engineering, 2012*(1), 1–15.

Feig , E., & Winograd, S. (1992). Fast algorithms for the discrete cosine transform. *IEEE Transactions on Signal Processing, 40*(9), 2174–2193.

Firmansah, L., & Setiawan, E. (2016). Data audio compression lossless FLAC format to lossy audio MP3 format with Huffman Shift Coding algorithm. *In Proc. of 4th International Conference on Information and Communication Technology (ICoICT)* (pp. 1–5). Bandung, Indonesia: IEEE.

Fraunhofer HHI. (March, 2017). Specifications for High-density Camera Array (HDCA) Data Sets. *Doc. ISO/IEC JTC1/SC29/WG1/M75008, 75th JPEG Meeting*. Sydney, Australia.

Gillian, C., Dragotti, P., & Brookes, M. (2014, Feb.). On the spectrum of the plenoptic function. *IEEE Transactions on Image Processing, 23*(2), 502–516.

Goodman, J. W. (2004). *Introduction to Fourier optics*. Greenwood Village, USA: Roberts & Company.

Google. (2019, Feb. 11). *A New Image Format for the Web*. Retrieved June 10, 2019, from WebP: https://developers.google.com/speed/webp/

Grossmann, A., & Morlet, J. (1984). Decomposition of Hardy functions into square integrable wavelets of consttant shape. *SIAM J. Math. Anal,, 15*, 723–736.

Haddad, R., & Akansu, A. (2012). *Multiresolution Signal Decomposition: Transforms, Subbands, and Wavelets*. San Diego, California: Academic Press.

Han, J., Saxena, A., & Rose, A. (2010). Towards jointly optimal spatial prediction and adaptive transform in video/image coding. *Proc. of IEEE Intl. Conf. on Acoustics, Speech and Signal Processing*, (pp. 726–729).

Hanhart, P., Bernardo, M. V., Pereira, M., Pinheiro, M. V., & and Ebrahimi, T. (2015, Dec.). Benchmarking of objective quality metrics for HDR image quality assessment. *EURASIP Journal on Image and Video Processing, 2015:15*, 1–18.

Hass, C. (2018, Jan.). *JPEG Compression Quality from Quantization Tables*. Retrieved June 2019, from Impulse Adventure: https://www.impulseadventure.com/photo/jpeg-quantization.html

HCI 4D Light Field Dataset. (2021). Retrieved from https://lightfield-analysis.uni-konstanz.de/

Hiroyuki, K., & Hitoshi, K. (2017). An Extension of JPEG XT with JPEG2000. *IEEE International Conference on Consumer Electronics (ICCE)* (pp. 373–374). Taiwan: IEEE.

Hoffman, D., Girshick, A., Akeley, K., & Banks, M. (2008, March). Vergence-accommodation accommodation conflicts hinder visual performance and cause visual fatigue. *Journal of vision, 8*(3), 114–130.

Hou, H. (1987, Oct.). A fast recursive algorithm for computing the discrete cosine transform. *IEEE Transactions on Acoustics, Speech, and Signal Processing., 35*(10), 1455–1461.

Howard, P. (1996). Lossless and lossy compression of text images by soft pattern matching. *In Proc. of Data Compression Conference*, (pp. 210–219). Snowbird, UTAH.

Howard, P. G., & Vitter, J. C. (1992, Dec.). Analysis of arithmetic coding for data compression. *Information Processing and Management, 28*(6), 749–763.

Howard, P. G., & Vitter, J. S. (1993). Design and analysis of fast text compression based on quasi-arithmetic coding. *in Proc. Data Compression Conference*, (pp. 98–107). Snowbird, UTA.

Howard, P., Kossentini, F., Marins, B., Forchhammer, S., & Rucklidge, W. (1998, Nov.). The emerging JBIG2 standard. *IEEE Transactions on Circuits and Systems for Video Technology, 8*(7), 838–848.

Huffman, D. A. (1952). A method for the construction of minimum redundancy. *Proc. of IRE, 40*, pp. 1098–1101.

Huffman, D. C. (2008). Vergence-accommodation conflicts hinder visual performance and. *J. Vis., 8*(3), 33–33.

ISO/IEC JTC 1/SC 29/WG1 N88015. (July 2020). *JPEG Pleno Holography Uses Cases and Requirements.* Informative distribution WG1, Geneva. Retrieved Dec. 20, 2020

ISO/IEC JTC 1/SC29/WG1 N88014. (2020). *Final Call for Evidence on JPEG Pleno Point Cloud Coding.* Retrieved Dec. 20, 2020

ISO/IEC JTC1/SC29/WG11 MPEG2016/n18665. (2019). *Common Test Conditions for.*

ISO/IEC-10918. (1994). Information technology -- Digital compression and coding of continuous-tone still images: Requirements and guidelines.

ISO/IEC-14495. (1999). Information technology -- Lossless and near-lossless compression of continuous-tone still images.

ISO/IEC-14496-12. (2015, Dec.). Information Technology - Coding of Audio - Visual Objects -- Part 12: ISO Base Media File Format.

ISO/IEC-15444-1. (2004). Information technology -- JPEG 2000 image coding system: Core coding system.

ISO/IEC-15444-12. (n.d.). Information technology: JPEG 2000 image coding systems: ISO base media file format.

ISO/IEC-18477. (Under publication). Information technology -- Scalable compression and coding of continuous-tone still images.

ISO/IEC-24800-1:2012. (2012). *Information technology - JPSearch - Part 1: System framework and components.* ISO/IEC. Retrieved April 8, 2019, from ISO/IEC: https://www.iso.org/standard/60232.html

ISO/IEC-24800-2:2011. (2011). *Information Technology - JPSearch - Part 2: Registration, identification and management of schema and ontology.* ISO/IEC. Retrieved Sept. 2019, from https://www.iso.org/standard/50411.html

ISO/IEC-24800-3:2010. (2015, May). *Information technology - JPSearch - Part 3: Query format.* ISO/IEC. Retrieved Sept. 2019, from https://www.iso.org/standard/50412.html

ISO/IEC-24800-4:2010. (2010). *Information Technology - JPSearch - Part 4: File format for metadata embedded in image data (JPEG and JPEG 2000).* ISO/IEC. Retrieved Sept. 2019, from https://www.iso.org/standard/50413.html

ISO/IEC-24800-5:2011. (2011). *Information Technology - JPSearch - Part 5: Data interchange format between image repositories.* ISO/IEC. Retrieved Sept. 2019, from https://www.iso.org/standard/50414.html

ISO/IEC-29170-2. (2015, Aug.). Information technology -- Advanced image coding and evaluation -- Part 2: Evaluation procedure for nearly lossless coding.

ISO/IEC-29170-2:2015. (2018, July 17). *Information technology – Advanced image coding and evaluation – Part 2: Evaluation procedure for nearly lossless coding.* Retrieved from International Organization for Standardization: https://www.iso.org/standard/66094.html.

ISO/IEC-29199. (2010). Information technology -- JPEG XR image coding system.

ISO/IEC-DIS-21122. (Under development). Information technology -- JPEG XS low-latency lightweight image coding system -- Part 1: Core coding system.

ISO/IEC-JTC-1/SC-29/WG1. (2015). *JPEG Pleno - Scope, use cases and requirements, Ver 0.1.* Poland: International Organisation for Standardisation.

ISO/IEC--JTC1/SC29/WG1. (2018). *Final Call for Proposals for a Next-Generation Image Coding Standard (JPEG XL).* 79th Meeting, La Jolla, CA, USA, 9–15 April 2018: JPEG.

ISO/IEC--JTC1/SC29/WG1. (2018). *JPEG Pleno Holography--Overview of Holography 2.0, WG1N80042, 80th JPEG Meeting.* Berlin, Germany.

ISO/IEC-JTC1/SC29/WG1. (2018). *JPEG Pleno Point Clouds—Use Cases and Requirements WG1N80018, 80th JPEG meeting.* Berlin, Germany, July 2018.

ISO/IECJTC1/SC29/WG1/N84025. (2019). JPEG Pleno light field coding common test conditions V3.3. *84th Meeting*, (pp. 1–26). Brussels, Belgium.

ISO/IEC-JTC-1-SC29. (2018, Oct. 30). *JPEG Pleno (Plenoptic image coding system).* Retrieved from Programme of work: https://www.itscj.ipsj.or.jp/sc29/29w42901.htm#JpegPLENO

ISO/IET JTC1/SC29/WG11, & WG11N18189. (2019). *G-PCC codec description v2.*

ISO/IET JTC1/SC29/WG11, & WG11N18475. (2019). *"V-PCC Test Model v6.*

ISO-23008-2-ITU-T-Rec.H.265. (2018). *High efficiency video coding.*

ITU-R. (1990). CCIR Rec. 601-2. *Encoding parameters of digital television for studios.*

ITU-R-BT.500-13. (2012). *Methodology for the subjective assessment of the quality of television pictures.* International Telecommunication Union. Retrieved Jan. 2012

ITU-R-BT.709-6. (2015). *Parameter values for the HDTV standards for production and international programme exchange.* International Telecommunication Union.

ITU-T-P.910. (April 2008). *Subjective video quality assessment methods for multimedia applications.* Retrieved Jun 10, 2019, from https://www.itu. int/rec/T-REC-P.910-200804-I

ITU-T-Rec-BT.2100. (2017). *Image parameter values for high dynamic range television for use in production and international programme exchange.*

ITU-T-T.809. (2011, May). Information technology JPEG 2000 image coding system: Extensions for three-dimensional data, SERIES T: Terminals for telematic services still-image compression JPEG 2000. ITU-T.

ITU-T-T.82/ISO-IEC-11544. (1993). *Information technology - Coded representations of pictures and audio information - Progresive bi-level image compression.* International Telecommunication Union. Retrieved Feb. 18, 2019, from https://jpeg.org/jbig/workplan.html

ITU-T-T.832/ISO-IEC-29199-2. (2009). Information technology—JPEG XR image coding system—Image coding specification, ITU-T CCITT Recommendation, 2009.

ITU-T-T.88/ISO-IEC-14492. (2018, Aug.). *Information technology – Lossy/ lossless coding of bi-level images.* International Telecommunication Union - International Organization for Standardization and the International Electrotechnical Commission.

Iwahashi, M., & Kiya, H. (2012). Efficient lossless bit depth scalable coding for HDR images. *Proc. of Asia Pacific Signal and Information Processing Association Annual Summit and Conference*, (pp. 1–4). Hollywood, CA.

Iwahashi, M., & Kiya, H. (2013). Two layer lossless coding of HDR images. *IEEE Intl. Conf. on Acoustics, Speech and Signal Processing (ICASSP)* , (pp. 1340–1344). Vancouver, BC.

Jack, K. (2008, Jan.). *Digital Video and DSP.* Elsevier. doi:https://doi. org/10.1016/B978-0-7506-8975-5.X0001-5

Jain, A. K. (1989). *Fundamentals of digital image processing.* Engleewood Cliffs: Prentice Hall.

JPEG. (2017, Jan. 20). *JPEG Pleno Database: 8i Voxelized Full Bodies (8iVFB v2) - A Dynamic Voxelized Point Cloud Dataset.* Retrieved Dec. 20, 2020, from JPEG Pleno Database: http://plenodb.jpeg.org/ pc/8ilabs/

JPEG. (2018, April). *Documentation on JPEG AIC.* Retrieved 2019, from JPEG: https://jpeg.org/aic/documentation.html

JPEG. (2018, April). *Documentation on JPEG Pleno.* Retrieved 2019, from JPEG: https://jpeg.org/jpegpleno/documentation.html

JPEG. (2018, April). *JPEG Pleno Database*. Retrieved 2019, from JPEG: https://jpeg.org/jpegpleno/plenodb.html

JPEG. (2018, April). *JPEG Pleno Holography*. Retrieved 2019, from JPEG: https://jpeg.org/jpegpleno/holography.html

JPEG. (2018, April). *JPEG Pleno Light Field*. Retrieved July 2019, from JPEG: https://jpeg.org/jpegpleno/lightfield.html

JPEG. (2018, April). *JPEG Pleno Point Cloud*. Retrieved 2019, from JPEG: https://jpeg.org/jpegpleno/pointcloud.html

JPEG. (2018, April). *JPEG-LS Software*. Retrieved 2019, from JPEG: https://jpeg.org/jpegls/software.html

JPEG. (2018, Oct. 30). *Overview of JPEG*. Retrieved 2018, from JPEG: https://jpeg.org/jpeg/index.html

JPEG. (2018, Nov.). *Overview of JPEG*. Retrieved 2018, from https://jpeg.org/jpeg/index.html

JPEG. (2018, Nov.). *Overview of JPEG*. Retrieved 2018, from JPEG: https://jpeg.org/jpeg/index.html

JPEG. (2018, Oct. 30). *Overview of JPEG 2000*. Retrieved from JPEG: https://jpeg.org/jpeg2000/index.html

JPEG. (2018). *Overview of JPEG Systems*. Retrieved Sept. 2019, from JPEG: https://www.itscj.ipsj.or.jp/sc29/29w42901.htm#JPEGXT

JPEG. (2018, Oct. 30). *Overview of JPEG XR*. Retrieved 2019, from JPEG: https://jpeg.org/jpegxr/index.html

JPEG. (2018, Oct. 30). *Overview of JPEG XS*. Retrieved 2019, from JPEG: https://jpeg.org/jpegxs/index.html

JPEG. (2018, Oct. 30). *Overview of JPEG XT*. Retrieved from JPEG: https://jpeg.org/jpegxt/index.html

JPEG. (2018, Oct. 30). *Overview of JPEG-LS*. Retrieved 2019, from JPEG: https://jpeg.org/jpegls/index.html

JPEG. (2018). *Overview of JPSearch*. Retrieved April 3, 2019, from JPEG: https://jpeg.org/jpsearch/index.html

JPEG. (2018, April). *Workplan & Specs of JPEG AIC*. Retrieved from JPEG: https://jpeg.org/aic/workplan.html

JPEG. (2018, April). *Workplan & Specs of JPEG Pleno*. Retrieved 2019, from JPEG: https://jpeg.org/jpegpleno/workplan.html

JPEG. (2018). *Workplan & Specs of JPEG-LS*. Retrieved 2019, from JPEG: https://jpeg.org/jpegls/workplan.html

JPEG. (2019, July 26). *Overview of JBIG*. Retrieved July 26, 2019, from JPEG: https://jpeg.org/jbig/index.html

JPEG. (2019). *Overview of JPEG AIC*. Retrieved from JPEG: https://jpeg.org/aic/index.html

JPEG. (2019, April 15). *Overview of JPEG Pleno*. Retrieved from https://jpeg.org/jpegpleno/index.html

JPEG. (2019). *Overview of JPEG XL*. Retrieved April 22, 2019, from JPEG: https://jpeg.org/jpegxl/index.html

JPEG Pleno. (2019). *Study text for DIS of 21794-2 (JPEG Pleno Part 2), Doc. ISO/IEC JTC 1/SC*. Geneva, Switzerland.

JPSearch-RA. (2011). *JPSearch Core Metadata Schema*. Retrieved April 2, 2019, from http://dmag5.pc.ac.upc.edu:8080/jpsearch-ra/jpsearch_coreschema.jsp

Kim, M. K. (2011). *Digital Holographic Microscopy: Principles, Techniques, and Applications*. New York, USA: Springer Series in Optical Sciences.

Kohn , M. (2005, Aug.). *Huffman/CCITT Compression In TIFF*. Retrieved July 2019, from Mike Kohn's Website: https://www.mikekohn.net/file_formats/tiff.php

Kuang, J., Johnson, G. M., & Fairchild, M. D. (2007). iCAM06: A refined image appearance model for HDR image rendering. *Journal of Visual Communication and Visual Representation, 18*, 406–414.

Kumar, B. V., & Naresh,, P. S. (2016). Generation of JPEG quantization table using real coded quantum genetic algorithm. *Intl. Conf. on Communication and Signal Processing (ICCSP)*, (pp. 1705–1709). Melmaruvathur.

Langdon, G. G. (1984, March). An introduction to arithmetic coding. *IBM Journal of Research and Development, 28*(2), 135–149.

Lee, B. (1984, Dec.). A new algorithm to compute the discrete cosine transform. *Speech, and Signal Processing, 32*(6), 1243–1245.

Leong, M., Chang, W., Houchin, S., & Dufaux, F. (2007). *JPSearch—Part 1: system framework and components. ISO/IEC JTC1/SC29 WG1 TR 24800-1:2007*.

Levoy, M., & Hanrahan, P. (1996). Light field rendering. *Proc. of ACM 23rd Intl. Conf. on Comp. Graphics and Iterative Techniques (SIGGRAPH)*, (pp. 31–42). New Orleans, LA.

Li, Z., Aaron, A., Katsavounidis, I., Moorthy, A., & Manohara, M. (2016, Jun 05). *Towards a practical perceptual video quality metric*. Retrieved Jun 10, 2019, from Netflix Technology Blog: https://medium.com/netflix-techblog/toward-a-practical-perceptual-video-quality-metric-653f208b9652

Loeffler, C., Ligtenberg, A., & Moschytz, G. (1989). Practical fast 1-D DCT algorithms with 11 multiplications. (pp. 988–991). Proc. of Intl. Cof. on Acoustics, Speech, and Signal Processing.

Lohmann, A. W., & et al. (1996). Space-bandwidth product of optical signals and systems. *J. Opt., A13*(3), 470–473.

Long- Wen , C., Ching- Yang , W., & Shiuh-Ming , L. (1999). Designing JPEG quantization tables based on human visual system. *Proceedings of Intl. Conf. on Image Processing (Cat. 99CH36348)*, (pp. 376 - 380). Kobe, Japan.

Loop, C., Cai, Q., Orts, S., & et al. (2016). *Microsoft Voxelized Upper Bodies – A Voxelized Point Cloud Dataset, ISO/IEC JTC1/SC29 Joint WG11/ WG1 (MPEG/JPEG) input document m38673/M72012*. Geneva, Italy.

LSoft-Technologies. (2019, Jan.). *JPEG 2000 Signature Format: Documentation & Recovery Example*. Retrieved Aug. 2019, from Active File Recovery: https://www.file-recovery.com/jp2-signature-format.htm

Malvar, H., & Sullivan, G. (2003). *YCoCg-R: A color space with RGB reversivility and low dynamic range*. Trondheim, Norway: Moving Picture Experts Group and Video Coding Experts Group.

Mannos, J. L., & Sakrison, D. J. (1974, July 4). The effects of a visual fidelity criterion of the encoding of images. *IEEE Transactions on Information Theory, 20*(4), 525–536.

Marcellin, M., Lepley, M., Bilgin, A., Flohr, T., Chinen, T., & Kasner, J. (2002, Jan.). An overview of quantization in JPEG 2000. *Signal Processing: Image Communication, 17*(1), 73–84.

Marints, B., & Forchhammer, S. (1997). Lossless/lossy compression of bi-level images. *In Proc. of IS&T/SPIE Symposium on Electronic Imaging Science and Technology, 3018*, pp. 38–49. San Francisco, California.

Marpe, D., Schwarz, H., & Wiegand, T. (2003, July). Context-based adaptive binary arithmetic coding in the H.264/AVC video compression standard. *IEEE Transactions on Circuits and Systems for Video Technology, 13*(7), 620–636.

Martucci, S. A. (1990). Reversible compression of HDTV images using median adaptive prediction and arithmetic coding. *In Proc. IEEE Intl. Symp. Circuits Syst.*, (pp. 1310–1313). New Orleans, LA.

Masaoka, K. (2016). Display Gamut Metrology Using Chromaticity Diagram. *IEEE Access, 4*, 3878–3886.

McKeefry, D. J., Murray, I. J., & Kulikowski, J. J. (2001, Jan.). Red–green and blue–yellow mechanisms are matched in sensitivity for temporal and spatial modulation. *Vision Research, 41*(2), 245–255.

Merhav, N., Seroussi, G., & Weinberger, M. (2000, Jan.). Coding of sources with two-sided geometric distributions and unknown parameters. *IEEE Transactions on Information Theory, 46*(1), 229–236.

Microsoft. (2018, 05 30). *HD Photo Format Overview*. Retrieved 08 16, 2019, from Microsoft: https://docs.microsoft.com/en-us/windows/win32/wic/ hdphoto-format-overview#encoding

MPEG, t. (1988). *MPEG-7.* Retrieved Sept. 2, 2019, from ISO/IEC JTC 1/SC 29/WG 11: https://mpeg.chiariglione.org/standards/mpeg-7

MPV. (2018). *MPV.* Retrieved from https://mpv.io

Müller, D.-I. K. (2019, July 07). *Dr.-Ing. Karsten Müller.* Retrieved from http://www.hhi.fraunhofer.de/ : http://iphome.hhi.de/mueller/index.htm

Netravali, A., & Limb, J. (1980). Picture coding: A review. *In Proc. of IEEE, 68*, pp. 366–406.

Ng, R., Levoy, M., Brédif, M., Duval, G., Horowitz, H., & Hanrahan, P. (2005). *Light field photography with a hand-held plenoptic camera.* Computer Science Technical Report.

Ngan, K. N., Leong, K. S., & Singh, H. (1989). Adaptive cosine transform coding of images in perceptual domain. *IEEE Transactions on Acoustics, Speech, and Signal Processing, 37*(11), 1743–1750.

Nill, N. B. (1985, June). A visual model weighted cosine transform for image compression and quality assessment. *IEEE Transactions on Communication, 33*(6), 551–557.

Nyquist, H. (1928, April). Certain topics in telegraph transmission theory. *Transactions of the American Institute of Electrical Engineers*, pp. 617–644.

Ochoa-Dominguez, H., & Rao, K. (2019). *Versatile Video Coding.* River Publishers.

Ono, F., Rucklidge, W., Arps, R., & Constantinescu, C. (2000). JBIG2-the ultimate bi-level image coding standard. *IEEE International conference on image processing (ICIP)* (pp. 140–143). Vancouver, BC, Canada: IEEE.

Oppenheim, A. V. (2014). *Discrete-time signal processing.* Pearson.

Pennebaker, W., & Mitchell, J. (1993). *JPEG, Still image data compression standard.* New York: Van Nostrand Reinhold.

Pereira, F., & da Silva, E. (2016). Efficient plenoptic imaging representation: Why do we need it? *IEEE International Conference on Multimedia and Expo (ICME)*, (pp. 1–6). Seattle, WA.

Rabbani, M., & Joshi, R. (2002, Jan.). An overview of the JPEG 2000 still image compression standard. *Signal Processing: Image Communication, 17*(1), 3–48.

Rane, S., & Sapiro, G. (2001, Oct.). Evaluation of JPEG-LS, the New Lossless and Evaluation of JPEG-LS, the New Lossless and for Compression of High-Resolution Elevation Data. *IEEE Transactions on Geoscience and Remote Sensing*, 2298 - 2306.

Rao, K. (2006, Feb. 1). Presentation. 1–21. Arlington, Texas, USA, Texas: DIP Lab.

Rao, K., & Yip, P. (1990). *Discrete cosine transform*. New York: Academic Press.

Raytrix. (2019, April 15). *3D light field camera technology: Enabling many new applications*. Retrieved from Raytrix: https://raytrix.de/

Reichelt, S., Haussler, R., Leister, N., Futterer, G., Stolle, H., & Schwerdtner, A. (2010). Holographic 3-d displays - electro-holography within the grasp of commercialization. In N. Costa, & A. Cartaxo, *Advances in Lasers and Electro Optics* (Vol. 1, p. Online). IntechOpen.

Reinhard, E., Stark, M., Shirley, P., & Ferwerda, J. (2002, July). Photographic tone reproduction for digital images. *ACM Transactions on Graphics (TOG), 21*(3), 267–276.

Řeřábek, M., & Ebrahimi, T. (2016). New light field image dataset. *8th International Conference on Quality of Multimedia Experience (QoMEX*, (pp. 1–2). Lisbon, Portugal.

Řeřábek, M., Yuan, L., Authier, L., & et al. (July, 2015). EPFL Light-Field Image Dataset. *69th JPEG Meeting*. Warsaw, Poland.

Richardson, I. (2010). *The H.264 Advanced Video Compression Standard* . Wiley.

Richter, T. (2013). On the standardization of the JPEG XT image compression. *Picture Coding Symposium (PCS)*, (pp. 37–40). San Jose, California.

Richter, T. (2018). Presentation in the University of Texas at Arlington: JPEG XT A New - Still Image Coding Standard on the Basis of ISO/IEC 10918-1/T.81. Arlington, Texas, USA.

Richter, T., Artusi, A., & Ebrahimi, T. (2016, July). JPEG XT: A new family of JPEG backward-compatible standards. *IEEE MultiMedia, 23*(3), 80–88.

Richter, T., Bruylants, T., Schelkens, P., & Ebrahimi, T. (2018). Presentation in The University of Texas at Arlington: The JPEG XT Suite of Standards: Status and Future Plans. Arlingtog, Texas, USA.

Richter, T., Keinert, J., Descampe, A., & Rouvroy, G. (2018). Entropy coding and entropy coding improvements of JPEG XS. *Data Compression Conference (DCC)* (pp. 87–96). Snowbird, UT: IEEE.

Rissanen, J. (1983). A universal prior for integers and estimation by minimum description. *Ann. Statist., 11*(2), 416–432.

Rissanen, J., & Langdon, G. G. (1979, March). Arithmetic coding. *IBM Journal of Research and Development, 23*(2), 149–162.

Said, A., & Pearlman, W. A. (1996, June). A new fast and efficient image codec based on set partitioning in hierarchical trees. *IEEE Transactions on Circuits System Video Technology, 6*(3), 243–250.

Santa-Cruz, D., Grosbois, R., & Ebrahimi, T. (2002, Jan.). JPEG 2000 performance evaluation and assessment. *Signal Processing: Image Communication, 17*(1), 113–130.

Saxena, A., & Fernandes, F. C. (2013). DCT/DST-based transform coding for intra prediction in image/video coding. *IEEE Transactions on Image Processing, 22*(10), 3974–3981.

Sayood, K. (2003). *Lossless compression handbook: A volume in communications, networking and multimedia.* Academic Press.

Schelkens, P., Munteanu, A., Tzannes, A., & Brilawn, C. (2006). JPEG2000 Part 10 - Volumetric data encoding. *Proc. of IEEE International Symposium on Circuits and Systems* (pp. 3877–3881). Island of Kos, Greece: IEEE.

Shannon, C. E. (1948, July). A mathematical theory of communications. *Bell System Technical Journal, 27*(3), 379–423.

Shapiro, J. M. (1993, Dec.). Embedded image coding using zero trees of wavelet coefficients. *IEEE Transactions on Signal Processing, 4*(12), 3445–3462.

Sheikh, H. R., & Bovik, A. C. (2006, January). Image information and visual quality. *IEEE Transactions on Image Processing, 15*(2), 430–444.

Sheikh, H., & Bovik, A. (2016, Jan). Image Information and Visual Quality. *IEEE Transations on Image Processing, 15*(2), 430–444.

Simard, P., Malvar, H., Rinker, J., & Renshaw, E. (2004). A foreground-background separation algorithm for image compression. *Data compression conference (DCC)* (pp. 498–507). Snowbird, UT, USA: IEEE.

SMPTE-ST-2022. (2007). Moving serial interfaces (ASI & SDI) to IP.

SMPTE-ST-2110. (2018). Professional media over managed IP networks standards suite.

Srinivasan, S., Tu, C., Zhou, Z., Ray, D., Regunathan, S., & Sullivan, G. (2007). *An introduction to the HD Photo technical design.* Microsoft Corporation.

Sullivan, J., Ray, L., & Miller, R. (1991, Jan.). Design of minimum visual modulation halftone patterns. *IEEE Transactions on Systems, Man and Cybernetics, 21*(1), 33–38.

Suzuki, T., & Yoshida, T. (2017, Dec.). Lower complexity lifting structures for hierarchical lapped transforms highly compatible with JPEG XR standard. *IEEE Transactions on Circuits and Systems for Video Technology, 27*(12), 2652–2660.

Sze, V., Budagavi, M., & Sullivan, G. (2014). *High Efficiency Video Coding (HEVC): Algorithms and Architectures.* Springer.

Taubman, D., & Marcellin, M. W. (2001). *JPEG 2000: Image compression fundamentals, practice and standards.* Boston, MA: Kluwer Academic Publishers.

Taubman, D., Naman, A., & Mathew, R. (2017). FBCOT: a fast block coding option for JPEG 2000. *Proc. of SPIE+Photonics, Applications of Digital Image Processing XL, 10396*, pp. 10396 - 10396 - 17. San Diego, CA.

Taubman, D., Ordentlich, E., Weinberger, M., & Seroussi, G. (2002, Jan.). Embedded block coding in JPEG 2000. *Signal Processing: Image Communication, 17*(1), 49–72.

Temmermans, F., Döller, M., Vanhamel, I., Jansen, B., Munteanu, A., & Schelkens, P. (2013, April). JPSearch: An answer to the lack of standardization in mobile image retrieval. *Signal Processing: Image Communication, 28*(4), 386–401.

Tompkins, D., & Kossentini, F. (1999). A fast segmentation algorithm for bi-level image compression using JBIG2. *IEEE International conference on image processing (ICIP)* (pp. 224–228). Kobe, Japan: IEEE.

Tran, T. D., Liang, J., & Tu, C. (2003, June). Lapped transform via time-domain pre- and post-filtering. *IEEE Transactions on Signal Processing, 51*(6), 1557–1571.

Tu, C., Srinivasan, S., Sullivan, G. J., Regunathan, S., & Malvar, H. S. (2008). Low-complexity hierarchical lapped transform for lossy-to-lossless. *Proc. of SPIE, 7073*, pp. 70730C-1 thru 70730C-12. San Diego, CA.

UCL, U. d. (2019, July 25). *OpenJPEG*. Retrieved Jan. 5, 2019, from OpenJPEG: https://www.openjpeg.org/

Vaish, V. (2008). *The (New) Stanford Light Field Archive*. Retrieved Jan. 6, 2021, from Computer Graphics Laboratory, Stanford University: http://lightfield.stanford.edu/

Van Bilsen, E. (2018). *Advanced Image Coding*. Retrieved from AIC: http://www.bilsen.com/aic/index.shtml

Veerla, R., Zhang, Z., & Rao, K. (2012, Jan). Advanced image coding and its comparison with various still image codecs. *American Journal of Signal Processing, 2*(5), 113–121.

VeryDOC Company. (2002). *What is JBIG2?* Retrieved Jan. 2019, from VeryDoc: https://www.verydoc.com/jbig2.html

Vest, C. M. (1979). *Holographic interferometry*. New York, USA: John Wiley and Sons.

Vetterli, M., & Nussbaumer, H. (1984). Simple FFT and DCT algorithms with reduced number of operations. *Signal processing, 6*(4), 267–278.

Vetterli, M.; Ligtenberg, A. (1986, Jan.). A discrete Fourier-cosine transform chip. *IEEE Journal on Selected Areas in Communications, SAC-4*(1), 49–61.

Wallace, G. (1992). The JPEG still picture compression standard. *IEEE Transactions on Consumer Electronics, 38*(1), xviii-xxxiv.

Wang, Z. (1984, Aug.). Fast algorithm for the W transform and for the discrete Fourier transform. *IEEE Transactions on Acoustics, Speech, and Signal Processing, 32*(4), 803–8016.

Wang, Z., Bovik, A. C., Sheikh, H. R., & et al. (2004, April). Image quality assessment: from error visibility to structural similarity. *IEEE Transactions on Image Processing, 13*(4), 600–612.

Warmerdam, F., & Rouault , E. (2019, July 20). *JP2KAK – JPEG-2000 (based on Kakadu).* Retrieved Jan. 3, 2019, from GDAL: https://gdal. org/drivers/raster/jp2kak.html

Watanabe, O., Kobayashi, H., & Kiya, H. (2018). Two-layer Lossless HDR coding considering histogram sparseness with backward compatibility to JPEG. *Proc. of Picture Coding Symposium (PCS)* (pp. 11–15). San Francisco, California: IEEE.

Weinberger, M. J., Rissanen, J., & Arps, R. B. (1996, April). Applications of universal context modeling to lossless compression of gray-scale images. *IEEE Transations on Image Processing, 5*, 575–586.

Weinberger, M., & Seroussi, G. (2000, Aug.). The LOCO-I Lossless Image Compression Algorithm: Principles and Standardization into JPEG-LS. *IEEE Transactions onImage Processing, 9*(8), 1309–1324.

Weinberger, M., Seroussi, G., & Saprio, G. (2000). The LOCO-I lossless image compression algorithm: principles and standardization into JPEG-LS. *IEEE Transactions on Image Processing, 9*(8), 1309–1324.

Wien, M. (2015). *High Efficiency Video Coding: Coding Tools and Specification.* Springer.

Willème, A., Mahmoudpour, S., Viola, I., Fliegel, K., Pospíšil, J., Ebrahimi, T., . . . Macq, B. (2018). Overview of the JPEG XS core coding system subjective evaluations. *Proc. of SPIE Optical Engineering + Applications. 107521M*, pp. 1–12. San Diego, California,: SPIE.

Winograd, S. (1978, Jan.). On computing the discrete Fourier transform. *Mathematics of computation, 32*(141), 175–199.

Witten, I. H., Neal, R. M., & Cleary, J. G. (1987, June). Arithmetic coding for data compression. *Communications of the ACM, 60*(6), 520–540.

Wu, X., Choi, W.-K., & Memon, N. D. (1998). Lossless interframe image compression via context modeling. *in Proc. 1998 Data Compression Conference*, (pp. 378–387). Snowbird, Utah, USA.

Xinfeng , Z., Shiqi , W., Ke, G., Weisi , L., Siwei , M., & Wen , G. (2017). Just-noticeable difference-based perceptual optimization for JPEG compression. *IEEE Signal Processing Letters, 24*(1), 96 - 100.

Xing, Y. (2015). *Méthodes de compression pour les données holographiques numériques, Ph.D. Tesis.* Paris: TELECOM ParisTech.

Yamaguchi, M. (2016). Light-field and holographic three-dimensional displays (Invited). *Journal of the Optical Society of America A, 33*(12), 2348–2364.

Yamaguchi, M., Ohyama, N., & Honda, T. (1992). Holographic three-dimensional printer: new. *Appl. Opt, 31*(2), 217–222.

Yaraş, F., Kang, H., & Onural, L. (2010, March). State of the art in holographic displays: A survey. *Journal of Display Technology, 6*(10), 443–454.

Ye, Y., & Cosman, P. (2003, Aug.). Fast and memory efficient text image. *IEEE Transactions on Image Processing, 12*(8), 944 - 956.

Yoon, K., Kim, Y., Park, J.-H., Delgado, J., Yamada, A., Dufaux, F., & Tous, R. (2012, Aug.). JPSearch: New international standard providing interoperable framework for image search and sharing. *Signal Processing: Image Communication, 27*(7), 709–721.

Zhang, J. (2013). *Dense point cloud extraction from oblique imagery, MSc Thesis.* Rochester: Rochester Institute of Technology.

Zhang, J., Fowler, J., Younan, N., & Liu, G. (2009). Evaluation of JP3D for lossy and lossless compression of hyperspectral imagery. *Proc. of IEEE Int. Geoscience and remote sensing symp. (IGARSS), 4*, pp. 474 - 477. Cape Town, USA.

Index

A

Arithmetic coder 67, 71, 95, 174, 177, 180

B

Binary arithmetic coding 5, 74, 177, 231–232, 234
Biorthogonality 168, 170
Bjøntegaard metric 215, 228

C

Chromaticity diagram 45, 47–48, 82
Coded block pattern 117
Color space 36–40, 42–44, 86, 145, 150, 191, 209, 217, 232, 236, 259
Cumulative density function 72
Common identifier 144
Coding of AC coefficients 92
Coding of DC coefficients 91
Color gamut 2, 45, 48, 137, 220, 258–260
Color space 36–40, 42–44, 86, 145, 150, 191, 209, 217, 232, 236, 259
Continuous time Fourier transform 11
Continuous wavelet transform 157

D

DCT 1, 3–4, 29, 33–35, 49–56, 58–62, 65, 69, 82, 85–89, 97–104, 112, 115, 138, 140–142, 147, 152, 232, 234, 257
Discrete cosine transform of type I 51
Discrte cosine transform of type II 51
Discrete cosine transform of type IV 51
Discrete Fourier transform 51
Differential pulse code modulation 91
DST 49, 53–54
Discrete sine transform of type VII 51
Discrete time Fourier transform 11
Discrete time wavelet transform 159
Discrete wavelet transform 1, 152, 158–159, 161–162

E

Embedded block coding with optimized truncation 152, 175
Entropy coding 4–5, 67, 90, 93, 102, 106, 113, 119–120, 133, 152, 154, 166, 175, 177, 180, 182, 186, 194, 235

F

Forward core transform 114
Forward discrete cosine
    transform 87
Fast Fourier transform 58, 81
Finite impulse response 168
Fixed-length code 120
Field of view 208, 211

G

Generic coding 286–287

H

Halftone coding 286–287
High dynamic range 2–4, 6, 132,
    135–136, 138, 260, 277
HDR images 132, 135–137
High efficiency video coding 53
Hierarchical lapped
    transform 62–63
Holographic imaging 2–3, 5, 203,
    206–207
Huffman coding 69, 81, 93–94, 97,
    144, 241, 279, 284
Human Visual System 4, 32–33,
    136, 141, 209, 213, 215

I

Irreversible color transform 44, 157
Inverse DCT 50, 58, 88
Inverse discrete wavelet
    transform 162
Independent JPEG Group 88
Image quantization 23
Image reconstruction 21, 143

J

Joint bi-level image experts
    group 6, 279
JPEG file interchange format 85–86
Just noticeable distortion 88

Joint Photographic Experts
    Group 1, 85
JPEG 2000 1–2, 4–5, 44, 107–108,
    112, 135, 141, 147, 149–153,
    157, 159, 170–173, 175, 177,
    180, 182–187, 190–193, 209,
    217–218, 231, 260, 263–266,
    271–273, 279
JPEG 1–7, 32–33, 35, 44, 64, 69,
    81, 85–88, 91, 95–99, 102–108,
    111–112, 115–117, 119–123,
    127–128, 130–133, 135–145,
    147–153, 157, 159, 170–173,
    175, 177, 180, 182–187, 189–
    196, 198–199, 203, 206–210,
    212–214, 217–219, 221–222,
    226–227, 231–232, 235–242,
    244–245, 251, 253, 257–258,
    260, 263–266, 270–273,
    277–280
JPEG AIC 5, 231
JPEG LS 1, 3, 235–236, 263
JPEG Pleno 3, 5, 203, 206–210,
    212–214, 218–219, 221,
    226–227
JPEG Systems 6, 210, 277–278
JPEG XR 1, 3–4, 64, 105–108,
    111–112, 116–117, 119–123,
    127–128, 130–133, 135, 147,
    263
JPEG XS 2–3, 5, 189–196, 198
JPEG XT 1, 3–4, 103, 135–136,
    138–145, 147–148, 263
JPEG XL 6, 257–258
JPSearch 6, 210, 263–268, 270–275

L

Layered zero coding 184
Light-field imaging 208
Lossless coding 4, 68, 85, 99,
    115–116, 136, 140–141, 147,

186–187, 196, 213, 225, 235, 238, 241–243, 245–247, 250, 254–255, 286

M

Macroblock 110–119, 121, 124–126, 132–133
Minimum coded units 87
Most probable symbol 71, 180
Most significant bits 120, 254
Motion JPEG 2000 151, 183
Motion JPEG XR 106, 127–128, 131
Mean squared errors 27
Modulation transfer function 32

N

Nyquist rate 11–12, 14, 24
Nyquist sampling theorem 12

O

One dimensional sampling 8
Optimal quantizer 28–30

P

Post-compression rate-distortion 176, 180
Picture coding symposium 290
Photo overlap transform 113, 115
Pixels-per-inch 7
Perfect reconstruction 170–171
Peak signal to noise ratio 214
Point cloud imaging 208

Q

Quality of service 182
Quantization 4–5, 13, 23–27, 29–33, 35–36, 49, 58, 69, 86, 88–90, 92, 102–104, 106, 115–117, 126, 140, 150, 154, 157, 166, 172–174, 183, 186,

190, 192–194, 232, 238, 242, 245, 257
Quantization error 26, 29, 32
Quality metrics 214, 217, 223, 227, 261

R

Rate-Distortion 33, 140, 176, 180, 183, 189, 193, 215, 222
Rate-distortion optimization 176, 180
ReGioN of interest marker 182
Refractive index 211
Reverisible color transform 44
Run-Length encoding 67
Region of interest 2, 5, 132, 149, 182–183, 186, 222, 267

S

Spatial light modulator 211
Signal to noise ratio 214
Still picture interchange file format 86
Structural SIMilarity index 215
Symbol coding 285–286

T

Test material 6, 198–199, 226, 260
Tiles 5, 110–114, 116–117, 121, 123–124, 126–127, 132–133, 151–152, 154–155, 184–185
Tiling 5, 86, 106, 108, 110–111, 127, 132, 147, 149, 154–156
The uniform quantizer 25, 88, 192

W

Wide color gamut 2, 260
Windows imaging component 105

Y

YYCbCr formats 45

About the Authors

K. R. Rao received the Ph. D. degree in electrical engineering from The University of New Mexico, Albuquerque in 1966. He is now working as a professor of electrical engineering in the University of Texas at Arlington, (UTA) Texas. He has published (coauthored) 17 books, some of which have been translated into Chinese, Japanese, Korean, Spanish and Russian. Also as e-books and paper back (Asian) editions. He has supervised 112 Masters and 31 doctoral students. He has published extensively and conducted tutorials/ workshops worldwide. He has been a visiting professor in National university of Singapore and electronics and telecommunications research institute (ETRI), Taejon, Korea. He has been a keynote speaker in many national and international conferences. He has been a consultant to academia, industry and research institutes. He has been an external examiner for several M.S. and Ph. D. students worldwide. He has been a reviewer of research proposals from Brazil, China, India, Korea, Singapore, Taiwan, Thailand and US. He was invited to review applications for recruitment and/or promotion of faculty in various Universities (US and abroad). He is an IEEE Fellow. He has been a member of the academy of distinguished scholars, UTA.

Humberto Ochoa Domínguez received the Ph. D. degree in electrical engineering from The University of Texas, Arlington in 2004. He is currently working as a professor in the Department of Computer and Electrical Engineering in the University of Ciudad Juárez, (UACJ) México. He has published extensively and is a reviewer for referred journals. He is the author of (coauthored) 4 books and is an IEEE member and a consultant for the academy and the industry. He has supervised several Masters and doctoral students and served as external examiner for M.S. and Ph. D. students. He is also a reviewer of research proposals from the industry.

Shreyanka Subbarayappa received her B.E. degree in Telecommunication Engineering from Ramaiah Institute of Technology, Bangalore, India in 2009. She received her M.S. and Ph.D. degree in Electrical Engineering from The University of Texas at Arlington, Texas, U.S.A. in 2012 and 2020. She is

315

currently serving as an assistant professor at Ramaiah University of Applied Sciences, Bangalore India in Electronics and Communication Engineering department since January 2018. Prior to this, she has had 5 years of research and industry experience, working for Intel Corporation in US, UK and India as a Team Lead for the graphics driver development towards Windows, Linux and Android operating systems. She has delivered various invited lectures on "Data Compression" nationally as well as internationally. She has also worked as a visiting faculty at University of Texas at Arlington for a period of one year from 2019 to 2020 taking masters level courses. She was hired as a 'Subject Expert' by Google LLC, U.S.A. for advising the attorney on patent litigation cases on video codecs algorithms. She has served as a reviewer for many masters and Ph.D. thesis. Her passion and research interests are in the field of Multimedia Processing, Video Pre/Post Processing, Video Codecs, Image Pre/Post Processing, Image Codecs and Audio Codecs